ONCE UPON A TIME IN SPACE

FROM AWARD-WINNING FILMMAKER
JAMES BLUEMEL

ONCE UPON A TIME IN SPACE

OUR JOURNEY INTO THE UNKNOWN

CONTENTS

6
INTRODUCTION

10
PART ONE:
THE NEW GENERATION

100
PART TWO:
LIVING IN SPACE

172
PART THREE:
THE NEW SPACERS

242
PART FOUR:
THE NEW SPACE RACE

INTRODUCTION

What can we gain by sailing to the Moon if we are not able to cross the abyss that separates us from ourselves.

THOMAS MERTON, 20th century monk and theologian

Since Yuri Gagarin became the first person to orbit the Earth in 1961, there have been fewer than 700 people who have followed him into space. That figure is about to rise dramatically. Our view of space, and our relationship with it, is changing, and life on Earth will never be the same again. Once more, there is talk of sending people back to the Moon, this time to build permanent settlements and then, after that, to establish human colonies on Mars. Nation states and private companies are focused on who will be the first to reach these other worlds and capitalise on potentially valuable new territories. Fifty-five years ago, Neil Armstrong became the first man to walk on the Moon, winning the Space Race for America. Today, a new Space Race has begun and this time, it will usher in a future in which more and more of us will live off-Earth, and access to the stars will no longer be for the pioneering few.

This book sets the stage for that leap we're about to take; it traces the first 65 years of human space exploration, told by the explorers themselves: the men, women, astronauts, cosmonauts and space tourists, who left this planet and ventured out into the unknown. These pioneers represent the first wave of human space travel. Now, we are on the cusp of a new space age, which will see more people than ever before going to space: the second wave of space exploration.

As a documentary filmmaker, I became interested in hearing directly from the pioneers who have led us to the stars – those who risked everything to fulfil their dreams and demonstrate humanity's potential beyond Earth. Through hours of interviews with astronauts and cosmonauts, I explored their motivations, sacrifices, and often obsessive desire to venture into space. My primary focus, however, was on the human elements of their stories: the excitement, heartache, fear, and joy they experienced on their journeys. While the engineering marvels and scientific experiments in zero gravity are undoubtedly amazing, this book delves deeper into something astronauts and cosmonauts don't often discuss: their feelings. This exploration became a fascinating journey, often leading me to aspects of space travel I had not expected.

While I was following this human story of space exploration, some of the most revealing insights came from those who had stayed behind on Earth, as their partners, parents or children made the journey to space. For the obsessive adventurer, the chance to experience weightlessness, or see our planet as a shining blue marble in the darkness of space, outweigh the risks of something going wrong. Loss of life is a realistic possibility and for all these pioneers, the experience is worth that risk. However, it can be very different for their families.

Anna Fisher became the first mother in space when her daughter Kristin was nine months old. Today, Kristin can recognise her mother's achievements, in terms both of what it must have meant to her personally to accomplish a lifelong ambition, and what her professionalism and dedication did to further feminist discussion on the inclusion of women as part of NASA's astronaut corps. After becoming a mother herself, the gamble that her own mother was willing to take, when she was just a baby, has taken on a new significance. But the Fisher family was fortunate, and Anna returned safely home after a successful mission. Zhenya Lazutkin was eight years old when her father Sasha nearly died in a fire on the Mir space station. The experience of overhearing snatched news reports, and seeing her mother's obvious concern, traumatised young Zhenya to the extent that she was eventually hospitalised. Sean McCool also agreed to an interview with me, to speak for the first time about his experience of losing his father when the Space Shuttle Columbia disintegrated on re-entering the atmosphere. A university student at the time, Sean had gathered with friends to watch his father return from a successful space mission, live on television. Unsure about what he was seeing, he telephoned his father's friend, astronaut Dan Tani, who told him to make his way to Houston, as the worst had happened. During our interview, Sean realised that because of the way he had bottled up the grief of losing his father, his own children today know little about their grandfather, Willie McCool.

The individual journeys that people go on to become astronauts are similarly filled with complex emotions, struggles and, in a lot of cases, a determination bordering on madness. For Ron McNair, growing up in the segregated rural south of 1960s America, the likelihood of breaking out of poverty and becoming the second African American ever to go to space, was impossible to predict. His achievement has been an inspiration for people across the world, and his foundation has raised millions of dollars to help underprivileged children enter higher education. Anousheh Ansari used to dream of floating in space; it was her refuge, growing up during the violent times of the Iranian revolution. Iran does not have a space programme, and her hopes of one day going to space herself had no purchase to transcend the realm of fantasy. Her dream appeared so unlikely that she learnt to keep it hidden from school friends to avoid ridicule. Eventually, Anousheh would make a huge amount of money in the communications business, and she was able to buy a seat on a Russian Soyuz rocket. She became the world's first female space tourist, or space participant as I should correctly call her. For years, looking up towards the stars had been her source of fascination and escape. When she eventually reached space and looked out of the windows of the International Space Station (ISS), she was surprised to find it was not the stars that drew her attention, but the beauty of the planet she had dreamed so long of leaving.

Astronauts learn that events on Earth don't stop just because they themselves are in space. The separation between the home planet and a new world, whether that's a space station, a lunar base, or a Martian settlement, is something to which humans will have to adapt, as off-Earth living becomes a reality. It is something the pioneers of the first wave of space exploration have had to learn for themselves. Advances in video communications have meant that astronauts have attended birthday parties, anniversaries and even births, all whilst floating 200 miles above the Earth's surface. But if something does go tragically wrong, there is no quick way of getting back. Astronaut Dan Tani discovered how remote he was on his second trip to the ISS. As he was preparing to sleep, an emergency call was patched to him from mission control. He was told that his mother had been killed in a car crash. Unable to return home until the mission was complete, Dan grieved in space. Looking down on Earth from high above, he could see the town he grew up in and where his mother had lived all her life, and where her funeral would now take place. As more people venture into space, this separation between the worlds will become increasingly common, but on the 19th December 2007, Dan Tani became the first human to have to process the death of a loved one from space.

Nothing has shaped human space exploration more than the collaboration between two countries that started this journey into the stars as sworn enemies. Since the early 1990s,

America and Russia have worked continually and largely harmoniously in space, despite whatever political situations may have developed back here on Earth. This partnership, that culminated in the arrival of the first US-Russian crew on the new ISS in 1998, has been the bedrock of human space flight for over a quarter of a century. The long-proclaimed hope for space to truly unify humanity, found its physical manifestation in a man-made space station, continually inhabited by astronauts of different nations, and seen by millions of humans back on Earth, as a small bright dot in the sky, circling our world approximately every 90 minutes.

Just as our relationship with space is about to change, so too is this state of unity which has anchored manned space exploration for so long. In 2022, Putin's invasion of Ukraine once again positioned Russia and America on opposing ideological sides. In turn this situation has stress-tested the relationship between the two countries on the ISS. It is notable that, for a while, the only manifestation of collaboration between the two nations was on the ISS, where, against the odds, cooperation persisted. However, the space station is now nearing the end of its life, and a planned deorbiting is set for around 2030. As I write this, in 2025, there are currently no plans to continue the working relationship between America and Russia after the ISS has been deorbited. How we will explore space in the future is unclear, but with new players entering the arena, namely China and India, the only thing we know for certain is that things will change.

As a species, we now stand on the precipice of this exciting future, more divided than united, less like one people from one planet and more like individuals racing to claim the spoils of space for themselves. How we progress into this future is unclear, but in this book we can at least hear from the people who have been to space and navigated the first wave of off-Earth exploration. I hope the wisdom they have gained from their unique perspectives will be listened to. The questions of how we should proceed into this brave new world, need to be asked now; the voices that can guide us in a more united direction are the voices of those who have already been there, and can tell us what happened to them, Once Upon a Time in Space...

PART ONE:
THE NEW GENERATION

'We were bulletproof, we were immortal, we were astronauts.'

MIKE MULLANE, Astronaut

Quote from *Riding Rockets* by Mike Mullane (Scribner, 2006).

INTRODUCTION

A sense of renewed ambition filled the air in 1976 as NASA announced that they were looking for a new class of astronauts. The announcement signified a new era for human space exploration. Not only was America sending humans back into space, but they were also sending them on a pioneering new spacecraft, the Space Shuttle. Unlike anything that preceded it, the Space Shuttle would be the world's first reusable rocket, with the ambition of making human space flight faster and cheaper. The hope was that going to space would no longer need to be an exceptional event; it would become routine, cementing America's status as a space leader and proving their technical superiority on the world stage.

The announcement also included a new development with implications that would forever change the nature of space travel. Up until this point, NASA astronauts had been selected almost exclusively from the military, and all were white men. In 1978, that changed. NASA announced that it was looking to recruit civilian astronauts, with the selection open to anyone with a scientific or medical background, and that included women and people of colour. Astronauts would finally start to resemble the make-up of the country. This was indeed a forward-looking NASA, with plans for the world's most hi-tech spaceship and a modern, contemporary crew to fly her.

These new astronauts would come to be seen as trailblazers, not just in terms of operating a new space vehicle but in breaking societal barriers. Seeing women perform equally to their male counterparts challenged the chauvinistic belief that had declared women incapable of going to space. For many under-represented or suppressed groups in America, the importance of having visible role models in positions of power, responsibility, and acclaim was desperately needed. Seeing someone who looked like you, sounded like you, came from the same place as you, doing something inspirational, could be a powerful and effective way to broaden young people's horizons and expectations. The first African American astronauts were symbolic in that way.

Charles Bolden grew up in segregation in South Carolina and had already overcome huge hurdles to become a test pilot in the US Navy. One afternoon, a T38, a sleek two-seater jet used by NASA astronauts, landed on the airstrip of his base. Charlie watched as the young black man climbed down from the cockpit. Intrigued, he decided to introduce himself and find out who this guy was. Charlie didn't realise it then, but this meeting would alter the course of his life. Ron McNair was part of the new class of NASA astronauts, training to fly on the Shuttle. Like Charlie, he too had been raised in South Carolina and had experienced segregation and racism. Ron asked him if he would consider applying to NASA. Without thinking, Charlie responded that NASA would never take someone like him. Ron looked hard at him for a while before saying, "Well, that's the dumbest thing I've ever heard."

Charlie did apply to NASA and was accepted into the second class of Shuttle applicants. He would go on to fly four missions to space and eventually became the first African American to lead NASA. Ron McNair's fate sadly lay down a different path.

With America focusing on its Shuttle programme, the USSR followed its own path. Since they didn't put a man on the Moon and did not win the space race, they had pivoted to developing the world's first space station: a base in space for cosmonauts to live and work. Long-duration space missions were essential to master if the Soviets stood a chance of one day beating the Americans to Mars. For them, this was where the real prize was – not in regular flights to low-Earth orbit. Instead of improving access to lower Earth orbit with a new launch vehicle, the Soviets focused on understanding how humans responded to living in space, creating the world's first series of space stations. As America gazed inwards, enthralled by the success of the first Shuttle launches and confident in the superiority of their technology, the Soviet Union was quietly planning its own great leap forward. The competition that had powered the 1960s Space Race was still very much on. The stories of the individuals who lived through this transformative era, captured in their own voices, reveal the profound impact of these changes.

DREAMING OF SPACE

In 1976, 19 years after the launch of Sputnik, the world's first satellite, and a little less than four years after the last Apollo mission to the Moon, NASA announced that they were recruiting for a new class of astronauts that would better represent the diversity of modern-day America in terms of gender and race, in an attempt to rekindle the public's interest in space exploration. The successful candidates who were selected to make up NASA Astronaut Group 8 would be trained to fly NASA's new reusable spacecraft, the Space Shuttle. Then, in 1980, a further 19 astronauts were chosen to be part of NASA Astronaut Group 9 in order to supplement the 35 announced in 1978.

Above: NASA Astronaut Group 8, class photo, Teague Auditorium in January 1978.

Right top: Anna Fisher in 3rd grade, 1958.

Right bottom: 5 May 1961, Alan Shepard became the first American to travel to space, on a historic mission in the *Freedom 7* capsule.

Doctors Anna Tingle and Bill Fisher, who were engaged to be married, decided that they were going to apply for NASA Astronaut Group 8. For Anna, being an astronaut had been a lifelong dream.

ANNA FISHER: I was aware of Sputnik as a child, and I knew that the Russians had launched someone into space, but I didn't really follow the space programme at that time. I was always studying. It was really Alan Shepard's flight that motivated me.[1] Our teacher told us that the first American was going to launch into space, so we gathered around her transistor radio and listened to Shepard as he was talking to mission control. For some reason, it just captured my imagination, and I thought, *Wow, that would be so wonderful to combine my love of math and science with exploring.* It was at that moment I decided I wanted to be an astronaut.

Of course, it didn't seem like a very realistic goal. I was this shy, skinny kid who didn't look like the 'Right Stuff'.[2] I figured everybody would think I was nuts and was afraid people would laugh at me, so I never told anybody.

---

[1] Shepard became the first American to fly in space on 5 May 1961 when he piloted *Freedom 7* during the Mercury-Redstone 3 mission.
[2] The first intake of seven NASA astronauts was known as the 'Mercury Seven' or the 'Original Seven', and they were later featured in Tom Wolfe's critically acclaimed book *The Right Stuff*, the title of which referred to the mental and physical attributes that it was thought were needed to be a test pilot or astronaut.

THE NEW GENERATION

BILL FISHER: The word 'astronaut' wasn't coined until about 1958, so I wanted to be a spaceman. In the first grade, my teacher wrote to my parents that I would be a better student if I wasn't always 'living rockets'. My mother wrote back that she had the same problem at home, saying that I was always extremely 'rocket conscious'. It was all I wanted to do, and it stayed with me. When I was 12 years old, I applied to be one of the original seven astronauts. I still have the rejection letter, which essentially told me to go away and come back in about 20 years.

A fascination with space was something Mike Mullane, another future NASA Astronaut Group 8 candidate, also developed from a young age, particularly after the Soviet Union launched Sputnik.

MIKE MULLANE: Sputnik was launched on 4 October 1957 and started the space race. I remember waking up that morning and my dad was reading a paper, mad as hell. He had no idea what a satellite was; all he knew was the Russians had done it and we hadn't. He was so angry at Eisenhower for being asleep at the switch; for not having foreseen this and done it first, whatever 'it' was. This was in the deepest, darkest days of the Cold War, and Russia was our dreaded enemy, and now here they had launched this satellite. They were the evil empire.

After Sputnik was launched, one of the messages that was appearing in the papers was that America was growing soft, that we were turning out lawyers and other soft types of career fields instead of engineers and scientists. So, there was this huge push to get kids fascinated in science. We had a rocket club in high school where the chemistry teacher – and this is remarkable when you think about it – told us how to make some solid propellant rocket fuel. It was dangerous. It was explosive, but when you're a teenage boy, you don't have a brain in your skull, because it never crossed my mind that it was dangerous at the time, even though I would sit there and light a fuse and run and a rocket would blow up. I never thought, *This could be dangerous. I could die doing this.* My homemade rockets were pipe bombs with fins.

These were exciting days; everything was a first. And, frankly, Russia achieved most of these firsts. It was first after first after first. We were always behind them until maybe five years into it when we started catching up in a big way and began doing our own incredible things. But they were running away with it. We owe the 1969 landing on the Moon to Sputnik and to Russia. It's still amazing, when you think about it, that only 12 years after Sputnik, we had men on the Moon. I was all in.

THE NEW GENERATION

Los Angeles Times

VOL. LXXVI — IN FOUR PARTS — CC ★ — SATURDAY MORNING, OCTOBER 5, 1957 — 44 PAGES — PART 1 — ALL THE NEWS ALL THE TIME — DAILY 10¢

Polish Capital Swept by Riots
20,000 Battle Police and Militia, Shouting 'Down With Gomulka'

WARSAW, Oct. 4 (U.P.) — An estimated 20,000 Poles battled club-swinging police and militiamen tonight in Central Warsaw. Many of the rioters shouted for the downfall of Communist Party Chief Wladyslaw Gomulka.

Two separate clashes left a number of persons injured, including some women. It was the second straight night of rioting in the tense Polish capital.

Police Fire Tear Gas

An estimated 1000 steel-helmeted police and workers' militia charged with clubs and fired tear gas bombs to try to break up the rioting. The mobs fought back with their fists, rocks and paving blocks.

A United Press correspondent said members of the workers' militia pulled passengers from streetcars and beat them. He said he was among those beaten.

Rioting flared viciously for three hours and was still under way late tonight. Students surged into the streets and were joined by adults.

Pent-up Fury Breaks

Pent-up fury burst after police moved to break up a student protest rally called at the Polytechnic High School. The rally was called to condemn police treatment of students who demonstrated last night against the closing of the anti-Stalinist student newspaper Po Pro-

tles and catcalls against the government spread.

There were shouts of a kind never heard here before — "Down with Gomulka."

From the party building at a main intersection the rioting rampaged down adjacent side streets. Witnesses reported the police and militia used tear gas many times.

Action Defended

Police cordoned the Polytechnic High School and apparently several thousand students were caught inside. The secretariat of the party central committee, the top Communist body in Poland, said in a communique tonight that the student newspaper had "falsely and untruthfully presented the economic and political situation in the country, propagating views quite foreign to Socialism."

The communique said despite warnings from the central committee there was no improvement in the activity of the paper's editorial staff.

Riots Reveal Unrest, Says Exiled Leader

The latest riots in Poland reveal the general lack of freedom in that country, An-

Kuchel Gives Support to Knowland
First Major GOP Officeholder to Take Sides in Battle

BY ROBERT BLANCHARD

U.S. Sen. Kuchel, a 20-year-veteran of California political warfare, yesterday threw his unqualified support behind U.S. Sen. Knowland in his coming campaign for the Governorship.

In doing so, he became the first major Republican officeholder to take sides in the approaching battle between Knowland and Gov. Knight for the GOP nomination.

Kuchel, who carried on a vigorous campaign on his own behalf only last year, notified his Senatorial colleague of his support in a telegram addressed to Knowland at his home in Oakland.

Earned Respect

He pointed out he has known the senior Senator for more than 20 years and has served with him in both the State Legislature and the U.S. Senate.

He also noted Knowland is known for the respect he has earned among members of both parties in Congress.

"In my view," he declared, "you are highly and admirably qualified to serve the people of our great State as their chief executive with honor and distinction."

Russia Launches First Earth Satellite 560 Miles Into Sky

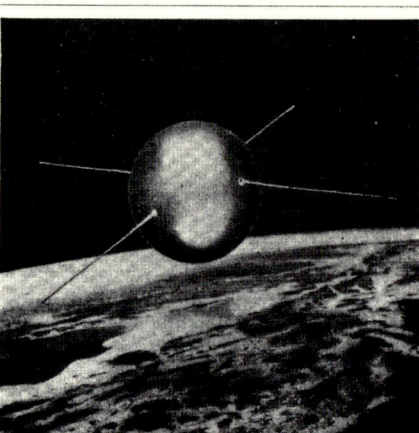

SATELLITE IN SKY—With Russian announcement that it has launched first earth satellite, this is artist's conception, superimposed on actual photograph of Southwestern United States from 140 miles up, of what the U.S. satellite would look like in flight. Drawing of 20-inch diameter globe was prepared at Naval Research Laboratory in Washington, D.C. Russian satellite is said to be 23 inches in diameter, weigh 185 pounds and orbit 560 miles above the earth.
U.S. Navy photo via Int. Wirephoto

'Moon' Carried Up by Multiple Rocket; Radio Signals Heard

MOSCOW, Oct. 5 (Saturday) (AP) — The Soviet Union announced today it has the world's first artificial moon streaking around the globe 560 miles out in space. A multiple-stage rocket launched the earth satellite yesterday, the Russians said, shooting it upward at about five miles a second.

They said the satellite, a globe described as 23 inches in diameter and weighing 185 pounds, can be seen in its orbit with glasses and followed by radio through instruments it carries.

In New York radio signals on the wave length of the Soviet moon—sounding as a deep "beep, beep, beep"—were picked up by electronic engineers of the National Broadcasting Co. The British Broadcasting Corp. in London reported similar reception.

Victory for Russia

In thus announcing the launching of the first earth satellite ever put in a globe-girdling orbit under man's controls, the Soviet Union claimed a victory over the United States. The two big powers had been in a hot but mainly secret race to be first to probe space with spheres laden with instruments.

The Moscow announcement said:

"The successful launching of the first man-made satellite makes a tremendous contribution to the treasure house of world science and culture . . ."

Space Travel Predicted

"Artificial earth satellites will pave the way for space travel and it seems that the present generation will witness how the freed and conscious labor of the people of the new Socialist society turns even the most daring of man's dreams into reality . . ."

In a special bulletin early this morning the Soviet Tass agency said the Russian moon "is now revolving around the earth at the rate of one circuit every hour and 35 minutes.

The launching occurred just three months and four days after the opening of the International Geophysical Year.

In Cambridge, Mass., officials at the Smithsonian Astrophysical Observatory said that sightings of the satellite had been reported last night by moonwatch stations across the country, including Whittier, Cal., Terre Haute, Ind., and Columbus, O.

Will Launch More

The observatory is the central point for collection of data on satellite observation by teams of "moon-watchers" in many parts of the world.

The Russian broadcast said the Soviet Union plans to launch several more earth satellites in the next year. It declared the development will open a way for travel to the planets.

Moscow said the satellite is fitted with steel radio transmitters continuously sending signals earthward on the 15 and 7.5-meter wave lengths and easily received by a broad range of amateur sets. Its announced weight of about 185 pounds is more than eight times the weight of a projected U.S. earth satellite.

Moscow described the signals as of about 3/10 of a second long with a pause of the same length. The two frequencies alternate in signalling, Moscow radio said. The broadcast said Soviet satellites planned later will

Turn to Page 4, Column 1

Left: Volunteers are trained as astronomers to watch satellites after the successful launch of Sputnik on 4 October 1957. This picture was taken in New York.

Right: Sputnik I, prior to its launch on 4 October 1957.

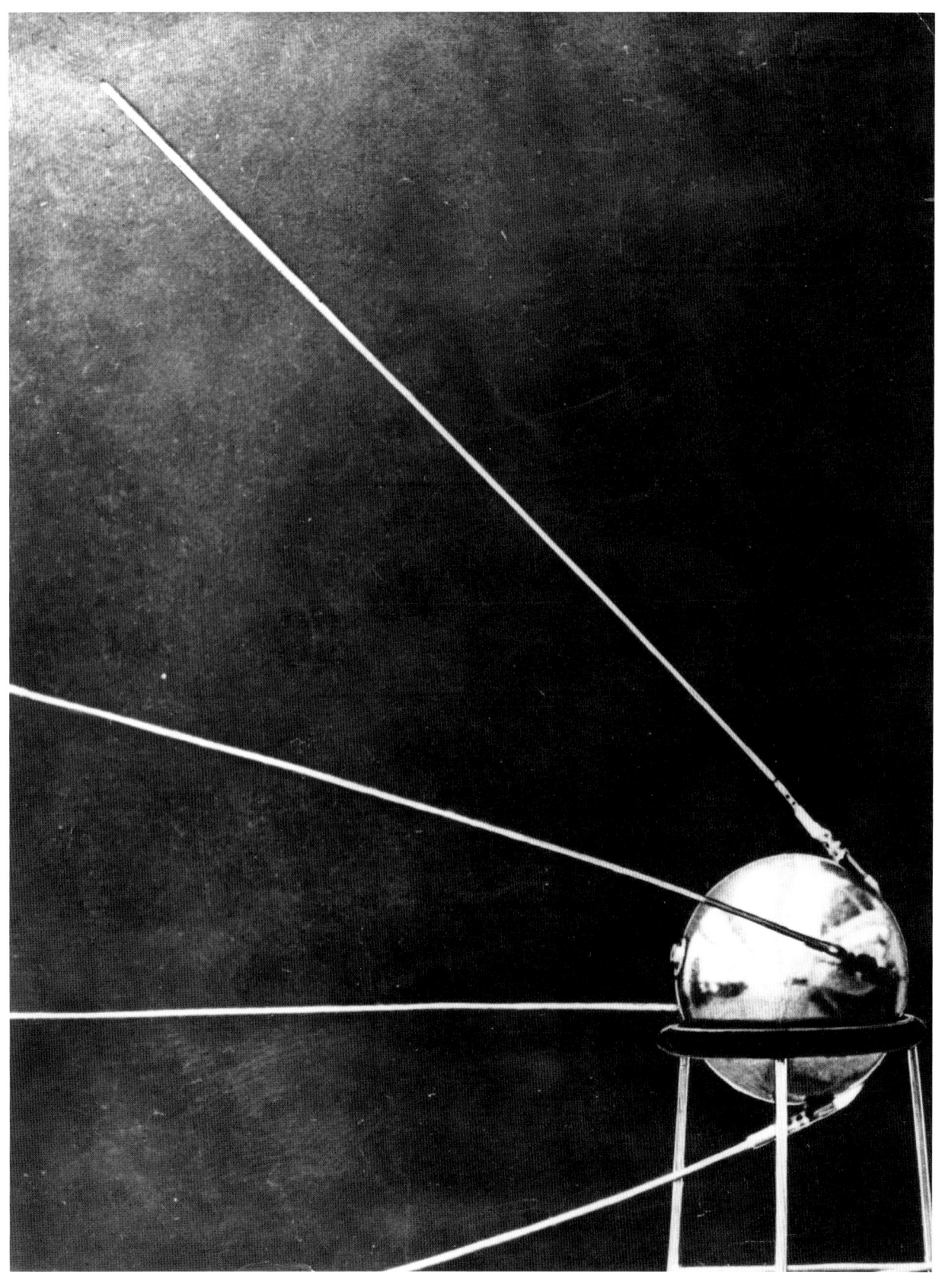

Meanwhile, hundreds of miles away in Costa Rica, Franklin Chang-Díaz, who was later chosen to be part of NASA Astronaut Group 9, also dreamed of becoming an astronaut as a child following the launch of Sputnik.

FRANKLIN CHANG-DÍAZ: I grew up at a time when there was a lot of change happening in the world. One of the events that had the biggest impact on me was the launch of Sputnik when I was seven years old. It was my mother who made me aware that this was a very important thing that had happened. The way my mom described it was that there was a new star that you could look at in the sky, a little point of light, and it was different from the others, because it moved. She said to me that if I looked at the sky in the early morning or the evening, I might be able to see it. Of course, I never really saw it, but in my mind I did, and I used to search for it. She also said that this star made by humans, the first one that had been launched, was the future.

I remember playing space explorers with my cousins and my friends. We even had our own spaceship. It was a cardboard box, and we had old radios and things like that so that we could pretend we were in a real spaceship. We would count down and launch into space and be home in time for dinner. A lot of my friends who were my former crew members who played with me in my cardboard box moved on to other more terrestrial pursuits, but I stayed with this idea that I wanted to be a space explorer. That was my calling. The typical reaction to this as a child growing up in Latin America was, 'Stop dreaming impossible dreams.'

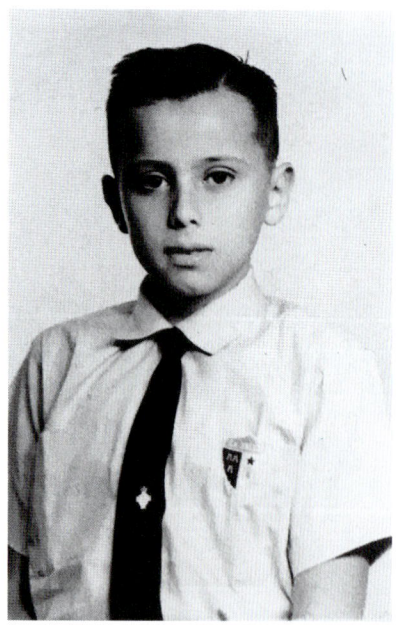

Below left: Franklin as a child in a toy car.
Below right: Franklin in the 1950s.

20 THE NEW GENERATION

Charles Bolden was also part of NASA Astronaut Group 9 and eventually rose to the highest office in NASA when he was appointed director of the agency by Barack Obama in 2009. Although his parents encouraged him when he was a child to follow his dreams, being an astronaut was not something he ever considered.

CHARLES BOLDEN: I grew up in Columbia, South Carolina. My mother and father were both schoolteachers. During that particular time in the Jim Crow South, segregation was the law of the land, so whites and Blacks didn't go to school together, didn't do anything together. We lived in different communities, and that's the way I grew up. I had fun, and life was good, but there were areas of the city that were white only. There were no signs that said STAY OUT but to get home we'd frequently end up running through the white neighbourhood being chased by some of the white kids. That was just life.

My parents were pretty good at telling me I could do anything I wanted to do, but also to be very careful: 'When you go out, don't be confrontational. If you're stopped by the police, do what they tell you to do, and just try to make it home safely.' That's 'the talk' that young Black men get even today.

Nevertheless, my mother and father always told us we could do anything we wanted to do. That we shouldn't let anybody else put limitations on us and we shouldn't put limitations on ourselves. My mother never, ever, gave up on anything.

Below left: Ethel Bolden and Charles F. Bolden, Senior, relax on their front porch, 1955.

Below right: Ethel and Charles F. Bolden, Senior, with Charles F. Bolden, Junior, and his brother Warren Bolden, 1964.

THE NEW GENERATION 21

Ronald McNair was selected to be an astronaut in NASA Astronaut Group 8 and became the second African American to fly in space. Ron was one of seven astronauts who died in the Space Shuttle Challenger disaster in 1986. His brother Carl remembers what it was like to be a child during segregation.

CARL McNAIR: Lake City was a typical small town in South Carolina during the 1960s. We knew our place as Black people. You didn't go across to the other side of the tracks, because that's where the white people lived. And, quite frankly, when we were growing up, it felt normal, because that's just the way it was. We had our own Black schools by law. We were pretty much second-class citizens.

You had 'white' water fountains, and you had 'coloured' water fountains, and we always thought there was something special about the 'white' water fountains, because why else were there two? 'Is that water better than our water?' Some of us got up the courage to take a sip, and I was one of them. It was just water. But for some reason or another, they considered us unclean.

Above: Ron McNair in elementary school in the 1950s.

Left: Young Black boy drinking at a water fountain with a 'Colored' sign above it in North Carolina in 1938.

Below: Actress Nichelle Nichols in *Star Trek* uniform,

CARL McNAIR: There were no Black people [going to space]. That wasn't our world. In 1969, we were still segregated. The only connection we had with space was *Star Trek*. That was it. We saw ourselves there thanks to Lieutenant Uhura. From the first time we saw her on screen, we were absolutely smitten. Oh, she just took our hearts. She was so beautiful. We'd never seen a Black woman on television. And not only was she a

Black woman, but she was an officer – from what I understand, fourth in command of a starship in the future. So, that told us that maybe there is a possibility for us to have a future in space. And I think Ron and I both saw that, though Ron took it a lot more seriously than the rest of us. However, there were some challenges. To begin with, all of the astronauts [at that time] were military test pilots, and they were all white men. Just a minor detail.

Below: A jeep and uniformed personnel are used to promote the TV show, *Men of Annapolis*, in Los Angeles, 1957.

CHARLES BOLDEN: At the age of 12, I saw a programme on television called *Men of Annapolis* about life at the Naval Academy. That was when I decided, That's where I'm going to school.

In order to get into college, you had to take the Scholastic Aptitude Test (SAT). You had to go to a particular testing centre to do it, and the only place where it was offered was the University of South Carolina, which was segregated at the time. When I got there that Saturday morning and they saw I was Black, they said, 'You can't test here.' I said, 'What do you mean I can't test here? I'm registered to test here.' So, they scrambled around for a little while and found a janitor's closet, and they put a little desk and a chair in there, and that's where I took my SAT. I was accepted to the University of Pennsylvania and Yale and then the Naval Academy, so it didn't keep me from doing well.

When I got there, although the Naval Academy was integrated, it was in a segregated city. After struggling to get to this place that I'd wanted to go to since I was 12 years old, I found out, oh man, people do not want me here! We started out with seven Black people in my class of about 1,400, and by the time we got to the end of my freshman year, we were down to four. I cried the whole first year, and my dad would tell me every single time that I called home, 'Hang in there one more week.' So, for 52 weeks, I hung in there one more week.

THE NEW GENERATION 23

ANNA FISHER: At the time I was growing up, your career options as a woman were basically to be a nurse, a teacher or a secretary. None of those interested me. Medicine seemed to me to be the best option, because, in the back of my mind, I thought, *If I don't get to be an astronaut, maybe I could be a doctor on a space station.* Again, this wasn't something that I told people about – it was just what was in my head.

When I was accepted into medical school in 1972, my class of 150 people at UCLA had only 16 women in it. There were times when a teacher would be showing slides – 'This is heart tissue. This is skin tissue' – and then, all of a sudden, a Playboy centrefold would flash up. It was totally inappropriate by today's standards, but that's what they were used to doing. I just said to myself, 'We're breaking into a male-dominated field, and this is the way things have been, so I'm not going to make a fuss. I'm just going to focus on my goals.'

I really loved medicine. I actually wanted to be a surgeon, and I applied and was accepted onto a surgery programme at a hospital that had never had a woman surgeon before. This was after the chairman of the department of surgery had told me that women did not belong in medicine, because they'd just leave when they had babies. One of the hardest things I ever did was to decide to not take that position, because I really wanted to show him he was wrong. However, that was when the chance to apply to be an astronaut came along.

Left: Anna working as a doctor in the late 1970s.

Below: Anna in her doctor's coat in the backyard, early 1970s.

Opposite: Wernher von Braun, the father of the US rocket programme in his office, 1964.

FRANKLIN CHANG-DÍAZ: Costa Rica had no space programme, and neither did Venezuela or any other country that I knew of. It was just the USA and the Soviet Union competing for supremacy, and we were all spectators. I recognised that I was not going to be an astronaut in Costa Rica, so I wrote a letter to another hero of mine, Wernher von Braun, who was the father of the US rocket programme. He had been brought over, along with several other German scientists, after World War Two, during which he had helped to develop the V-2 rocket, the world's first long-range ballistic missile that was used to bomb [the Allies on the Western Front]. He had been a member of the Nazi Party, so there was a lot of darkness in his history, but I took that out. To me, he was the foremost rocket scientist.

In my letter [to Braun], I said I was a kid from Costa Rica, that I lived down here, but now I wanted to go to the United States and be an astronaut and rocket scientist and work for NASA – what should I do? I wrote this letter in Spanish, of course, but somebody must have translated it or read it, because I got a response. Just getting an envelope from NASA [was amazing]. It took me a while to even open it, because I wanted to savour the moment. Inside was a form letter and somebody had taken the trouble to underline a sentence in the second paragraph with a red pencil that read, 'Careers with NASA are limited to United States citizens.' I didn't speak English, so I had the letter translated, including the bit that was underlined. However, I purposely misunderstood

the message. I interpreted it as them telling me to come to the United States and become a US citizen, and then they would give me a job. It seemed like a perfect invitation.

When I arrived in the fall of 1968, it was a very turbulent year in American history. Dr Martin Luther King Jr and Bobby Kennedy had both been assassinated. It was full of unrest, of racial tension. There were school fights, there were stabbings, and there were all kinds of issues and problems. Also, everything was fast. McDonald's was just beginning, and you could get a hamburger by just driving through.

Those were all culture shocks.

I convinced the high school officials through an interpreter to let me try and take the courses that a regular high school senior takes. I didn't speak the language yet, but I said I would learn it somehow. Amazingly enough – and to me this is one of the beautiful things about the United States – the school officials were progressive enough to say, 'Why not?'

One person in particular had a transformative impact on me, and that was Alan Winters. He was a teacher at Hartford High, and for some reason he took an interest in me. Every night, he would make me read pages from *A History of Western Philosophy* by Bertrand Russell. He said, 'Whatever you don't understand, here is a dictionary – look it up.' It forced me to really learn the language, and I was near the top of the class towards the end of the year, and I got a place at the University of Connecticut as a scholarship student.

While at university, Chang-Díaz witnessed a moment that loomed large in the minds of all of the future Group 8 and 9 astronauts.

Below: Buzz Aldrin's bootprint, one of the first steps taken on the Moon, from the Apollo 11 mission in July 1969.

Opposite: Buzz Aldrin, at the historic moment of the first extravehicular activity on the lunar surface, 1969.

FRANKLIN CHANG-DÍAZ: It was 20 July 1969. The school was not in session yet, but all the summer students were there, so the student union was pretty full of people to watch the [first] landing on the Moon. A TV set had been put out in one of the big rooms, and I was sitting at the front, watching every bit of the preparations for the landing, all the things that needed to be done. I knew everything that they were doing. I had learned the process of landing on the Moon, all the manoeuvres, and was very well informed. In fact, I was probably better informed than even [news anchor] Walter Cronkite, who was the guy who was narrating this to the American public. I noticed that there were people who were wondering how this Hispanic kid with a strange accent knows all this stuff and we don't.

When Neil Armstrong exited the LEM [Lunar Excursion Module] and began to move towards the lunar surface, you could hear a pin drop in the room. It was complete and total silence. When he jumped down the ladder and very carefully put [his foot] down on the surface, saying 'That's one small step for man, one giant leap for mankind,' everybody exploded. It was pandemonium. Some people were clapping, some people were crying. I was elated. I was watching those two astronauts hopscotching on the surface of the Moon on a little black-and-white TV screen, and I felt a certain kinship with them, as if my goal wasn't that far off any more.

BILL FISHER: I was teaching mountain climbing in Switzerland, and we were staying at an old hotel that had been converted into a mountain-climbing school. There were 12 of us: 11 teenagers and me, their instructor. We were training to climb Mont Blanc, but the date of our climb coincided with the lunar landing. Every single one of us voted not to climb and watch it instead, but we were told we couldn't leave the hotel. [The landing] was at something like 3am European time, so we all snuck out of the lodge and went down to the city of Lausanne, about half a mile away, and watched in a small bar. After the landing, lots of people were in the streets of Lausanne with torches, chanting, 'Sur la lune, sur la lune' – on the Moon. It was so spontaneous – a wonderful moment.

When I got back, I was fired. But the kids [who were with me] still remember it. One of the students later wrote a letter saying how much he appreciated being able to sneak out and watch the landing. I didn't regret a thing, although I would have liked to have stayed longer: mountain-climbing school wasn't half bad.

It wasn't something you could miss. I believe the first lunar landing was the most significant event in the history of mankind. Ten thousand years from now, if humans are still around, the name Neil Armstrong will be remembered.

Mike Mullane, meanwhile, was on a seemingly more traditional path by the standards of the time, heading towards a career in the military and marrying young.

Below: Mike and Donna on their wedding day.

MIKE MULLANE: Throughout high school, I had no real contact with any girls. I was afraid of them. Didn't know how to act around them. It was my sophomore year at West Point, and I was home for Christmas leave. This would have been 1964. There was alcohol at this party, and I started drinking to excess. My parents were there too, and my mom said, 'Michael, you're intoxicated. You need to get out and take a walk and get some air on you.' Donna followed me a few minutes later, and she talked to me briefly, and then she just grabbed me and kissed me. Out of nowhere. No girl had ever done that. I had never been kissed. When that happened, I was in heaven.

She gave me her address, so we basically fell in love through letters. Within six or eight months or so, I was mailing her an engagement ring. The rest is history. We got married a week after graduation, and almost immediately had two children, twins born nine months after we were married. At that point – I was 23 years old – I was wrapped up in my career. Donna's a trooper, and she stayed with me through a lot of tough times.

DONNA MULLANE: Mike was leaning against his car, and I just leaned over and kissed him, and that was it. He said, 'Let me have your address. I'll write to you.' I thought, *That'll never happen*. But it did. We were crazy kids because we got married and we had maybe been together a total of six to ten weeks. That's all the time we had spent with one another, because he was in New York and I was in Albuquerque. It is luck that we have been together for nearly 60 years. It really is.

Mike was obsessed with flying at this point. When he came from West Point to visit me, he'd get off the aeroplane holding a steel tube with fins on it. And what we would do during his Christmas break is go out in the desert and fire these rockets. How could I not have known? And from that time on, and especially when I became Mrs Mullane, it was a partnership, and his dream became my dream for him, and that's how we lived.

I was born into an Italian family and grew up Catholic. The priest would give you marriage classes, but what did they know about marriage? In fact, I still have my marriage book. It's a thick white scrapbook. The 1963 Marriage Course curriculum guide from St Mary's High School included a lesson on masculine and feminine psychology with a table of characteristics. Males are more realistic; females are more idealistic. Men are more emotionally stable; women are more emotionally liable. Man loves his work; woman loves her man. Men are more likely to be right; women are more likely to be wrong. How do you like that? That's what we were taught.

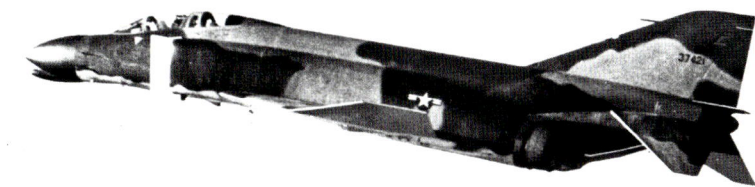

Right: An F-4 Phantom fighter jet, flying over Vietnam.

Below right: Mike Mullane in a RF-4 cockpit, during his time in the US Air Force.

MIKE MULLANE: We got married in 1967, and I was in Vietnam in 1969. This was a war against Communism, which we had all been taught was the greatest evil that existed at the time, and it was therefore something that we had to do. I know it's hard to believe that fighting and dying in Vietnam was somehow going to protect America, but that was the message: that Communism was a cancer that was going to grow and grow and grow and ultimately threaten the United States of America. So, I had no doubts about it, none at all. Looking back on it, it was a war we should have never been in, but, at the time, I couldn't see that. I wanted to go and do what those heroic airmen had done in World War Two – fight the enemy.

Thirty of my classmates out of West Point were killed over in Vietnam, so it was a grim time. I have to say, though, that I was much more frightened and scared when I was strapping into the Space Shuttle than I was when I was strapping into the cockpit over in Vietnam for a combat mission.

DONNA MULLANE: I knew that he might not come back. Television reports mainly showed the ground war, and it was always death and destruction, even though his letters to me would paint a picture of him coming home from a flight that had gone well and going to the officers' club to have a beer at seven o'clock in the morning because that was his dinner time. But what I saw on TV was what I know his friends and army buddies were going through. I knew planes were falling out of the sky as well.

He would call occasionally, but those calls were awful, because it was over the radio, so you'd have to say, 'I love you. Over.' 'How are the kids? Over.' 'The children are fine. Patrick's getting big. Amy is too. Over.' It was so impersonal.

It was wonderful when he came back, because he was safe, but he had changed a little. He wasn't as fun-loving as he had been prior to going. He eventually got back to being the old Mike, he really did, but you could tell that he'd had an experience that really touched him.

CHARLES BOLDEN: Becoming an astronaut wasn't something even remotely in my mind. I just wanted to get my wings and start flying my own aeroplane, which I did in 1970. The Vietnam War wasn't raging like it had been in the 1960s, but we were still at war.

I ended up going [to Vietnam] in the summer of 1972, and I stayed there a year. I don't know very many people who went to war and want to go back. Humans can be pretty violent. There were times when you weren't just attacking a building; but you could see people, and you knew that those people were part of your target, and you were going to play a part of ending their lives. That's not a pleasant thought.

In some ways, I felt like, *Why am I doing this?* I vehemently opposed the Vietnam War, but that was my job. I know that's very difficult for some people to understand. How can you do something that you think might be morally reprehensible? But I felt that I had an obligation to serve the nation. And, back then, we were told that if Vietnam fell, the rest of the world might fall. We went in to stop the reign of Communism that was coming over [Southeast Asia from China and the Soviet Union].

FRANKLIN CHANG-DÍAZ: I was a sophomore in the middle of my studies. The Vietnam War was going on, there were bombings in Cambodia, the Kent State shootings happened – lots of other things were going on in the country, and the Apollo programme got kind of lost in the noise of all the other stuff that was happening. My professors would tell me, 'Don't even think about joining NASA, because you're not going to get a job.' The space programme, which was really founded as a race against the Soviet Union, had accomplished its main objective, which was to beat the Russians to the Moon. And having accomplished that main objective, there was nothing coming behind it, so the programme was cancelled.

This was a lead brick that fell on the edifice of all of us who were looking to develop our careers in space. Thousands of aerospace engineers had lost their jobs. Many of them had graduated from the University of Connecticut, and I knew some of them. There was one guy from South Korea who ended up opening a gas station, and he had a PhD in aerospace engineering.

I had almost forgotten about the space programme. I was deeply involved in the beginnings of the rocket development of the VASIMR [Variable Specific Impulse Magnetoplasma Rocket] engine after being offered a job at the Charles Stark Draper Laboratory. These are the people who essentially wrote the book on how to get humans to the Moon. All of us at the Draper Lab were invited to watch a video [about the shuttle] called 'A Spaceship Landed on Earth; It Came from Rockwell' that was going to be released by Rockwell International. All of a sudden, the US space programme had woken up. It was back on. The sleepiness of the 1970s was now coming to an end, and a new programme was beginning.

NASA ASTRONAUT GROUP 8

It was against this backdrop of a seemingly dormant space programme that NASA announced that they were recruiting a new class of astronauts to take part in the Space Shuttle programme. Despite a successful career in medicine, Anna Fisher was still obsessed with space, and now she was engaged to someone who shared her passion.

ANNA FISHER: When I was a third-year medical student, I was getting a cup of coffee in the dining room, and apparently Bill saw me and was interested [in getting to know me]. Unlike me, he doesn't have a shy bone in his body – he was constantly telling us stories and cracking us up. One day, I was getting ready to go home, and I found a note on my car that said, 'I have a bottle of wine and some cheese. Why don't you stop by and say hi?'

One of the first things I liked about Bill was that we talked about space. He had wanted to be an astronaut since he was six years old, and I felt incredibly lucky that I had met someone who felt the same way I did. We were in the process of talking about those sorts of things when we found out that NASA was looking for a new type of astronaut.

When NASA called to ask if I was available to come to Houston for an interview, I remember putting my hand over the phone and saying to Bill, 'It's NASA. They want me to come for an interview, but it's the week that we are supposed to get married.' He replied, 'Just say yes. We'll figure it out.' So, I said, 'Yes, I can be there.' When I hung up, I said, 'Now what?'

It was a Friday, so we went to the Wayfarers Chapel, and they said they had an opening on the Tuesday. I ran out on Saturday and bought a dress, and we booked a photographer. We then had to tell everybody we were getting married that Tuesday. To this day, our friends still can't believe that shy, studious Anna had applied to be an astronaut. And I have to say, I still almost don't believe it myself. It just seems so surreal.

I remember thinking, *If I don't get picked, I don't know what I'm going to do. I won't ever feel the same about anything again, because this is what I really want.* For me, joining NASA was like going to Mecca.

My interview was on the Thursday right after my [NASA] eye exam, so my eyes were dilated, and everybody on the panel looked kind of fuzzy, which I think made things a bit easier. Everything went pretty well, and then at the end I said, 'Oh, by the way, just so you know, I really want to be an astronaut, but I do want to have a family, too.' To this day, I have no idea why I said that – I guess I felt like I should be honest about everything – but they picked me anyway. Later, I told my girls, 'When you go for a job interview, don't do what I did.'

BILL FISHER: In the time between receiving my rejection letter [when I was 12] and being accepted in 1980, I'd kind of given up on being an astronaut, because I wasn't a test pilot. I was always interested in medicine too, so I went to medical school instead.

Anna and I were eating lunch in the cafeteria at Harbor–UCLA Medical Center in June 1977. At that time, I was finishing my second year of general surgery residency, and Anna was completing her internship. A fellow physician was reading a magazine at the table, and in the course of our conversation, he happened to mention that it included an advert for an active astronaut selection process that was closing on 30 June 1977. It was 27 June.

We called NASA, and they sent us an application, which arrived on the 28th. There was only one overnight service at that time, so we filled out our applications and sent them back on the 29th, and they got there on the 30th. If we hadn't seen that magazine, we wouldn't be having this conversation.

MIKE MULLANE: I was at Edwards Air Force Base just finishing up with my test-pilot school class. This was probably June of 1976 or somewhere in that time frame, and the announcement that they were going to select astronauts for the shuttle programme appeared. This time, it wouldn't just be white males. They were going to have Black astronauts. They were going to have women astronauts.

As soon as I got home from work, I said to Donna, 'They announced they're selecting and I'm applying.' And, as she did throughout our whole life, she said, 'Go for it.' She was always supportive.

I didn't have my head in the sand. I knew it was a very long shot. Even to this day, I have no clue how I got selected. For the longest time, I thought it was a mistake, and they were going to realise at the last moment and pull me off the stage with a hook.

CARL McNAIR: Ron gave me a call one day and said, 'Hey, man, I don't know if I should tell you this, but I'm going to be an astronaut.' I said, 'You're going to be a what?' He said, 'I'm going to be an astronaut.' I said, 'How many people applied?' He said, 'I don't know, nine or ten thousand.' I said, 'How many astronauts are they looking for?' He said, 'They're looking for thirty-five.' I said, 'Let me get this right. Ten thousand people have applied, and they're looking for thirty-five astronauts. What makes you think you're going to be an astronaut?' And he said, 'Because I applied.' That's when I knew my brother had lost it. I said, 'Well, I'm going to be the Pope,' and thought, *We could play this game all day.*

A few weeks went by. [Then, on the day of the announcement,] the TV news reporter Walter Cronkite said, 'In from NASA, the first thirty-five space shuttle astronauts.' I thought, *Oh my god, this is going to hurt my brother's heart so bad.* Cronkite went down the list in alphabetical order. He got down to the 'M's, and then he said, 'Ron McNair.' I didn't hear another name. All I could think was, *He actually he did it. He actually did it.* I rushed to the phone: 'Congratulations, man, you did it.' Ron said, 'Did what?' I said, 'Man, you're an astronaut.' He said, 'I am? Let me call you back.'

I thought, *How is this happening?* He was about to take his own Starship *Enterprise* into space.

CHARLES BOLDEN: The women [from NASA Astronaut Group 8] appeared on the covers of *Life* and *Time* magazines. But if you wanted to read something about a Black person, you didn't go to *Time*. *Jet* and *Ebony* were focused on the Black community, and they both had Ron [McNair], Fred [Gregory] and Guy [Bluford] as their cover stories. Historically, many kids went to Black colleges and universities, but not many of them became astronauts. This was the first of a kind, and that was special. That got me. I was impressed.

Left: Official portrait of NASA astronaut class candidates (left to right) Ronald McNair, Guy Bluford and Fred Gregory at the Johnson Space Center in Houston, Texas, 1978.

Right: Front cover of *Jet* magazine in 1978.

THE NEW GENERATION 35

ANNA FISHER: I was just so excited. This was the dream I'd always had, and now I was going to get to live it. I was hoping that Bill was going to be picked too, but he found out that he wasn't selected. He was really good-natured about it. It was hard, though, because I wanted to be excited, but I didn't want to make Bill feel bad. He was asked lots of questions about how he felt, and all he ever said was, 'We're just glad one of us got accepted.' And he truly felt that way.

They brought all of us on the stage to introduce our class to the world. I never put anyone in our class in the same category as, say, the 'Mercury Seven', because that was the very beginning of humans ever leaving our planet, so, to me, they were true pioneers. I did realise that it was historically significant [that six women had been chosen], but I didn't spend a lot of time thinking about the historical perspective of it, other than being very happy that women were now being given the chance. I also thought to myself, *Don't screw up. We've got to make sure that all of us succeed so that the women who come after us will have the same opportunities.*

They did the formal announcement in the morning, and then in the afternoon, the women were separated from the guys, and they took pictures of the six of us together and individually. There was just so much interest in who we were and what our personal situations were. I think that made some of us women feel uncomfortable. People would ask us all these questions, but we just wanted to be part of the crew. I wanted the crew to share in the acknowledgement of the things we were doing.

Left: First women to be named by NASA as astronaut candidates (left to right) - Rhea Seddon, Anna Fisher, Judith Resnik, Shannon Lucid, Sally Ride and Kathryn Sullivan. Photographed at the Johnson Space Center, Houston, Texas, 1978.

Right: Anna and Bill with a shuttle model after Bill's selection, c.1980.

BILL FISHER: Anna and I were interviewed on a talk show in Los Angeles, and the host asked me, 'Bill, does it bother you that your wife got accepted and you didn't?' I said, 'No, I'm grateful Anna got in and that I got an interview.' And he said, 'You sure it doesn't bother you?' And I said, 'I'm sure.' And he said, 'But when you're alone driving your car, do you pound on the steering wheel?' But I really thought she was perfect for the job. There wasn't one iota of resentment or disappointment.

Before we left for Houston, I made a phone call to [Space Shuttle programme office manager] Mr Honeycutt[3] and said, 'Do me this favour: if I'm just not the guy, if you don't think there's any real chance, tell me so I can get on with my life. But if there is a chance, tell me what I need to do.' He said, 'Bill, we're really interested in you. You probably need to get another degree, but we're very interested. This is not a closed door.' Well, that's all I needed to hear.

So, we moved to Houston, and I got a degree in engineering in addition to my MD. Probably the most important thing I did, though, was play on the astronaut softball team. Astronauts could play and so could their spouses, and George Abbey [director of flight crew operations] was the coach. I wasn't the best player, but I could play a bit. And once, just once, I hit an in-the-park home run, and when I was coming around to our home plate, Mr Abbey was standing there to shake my hand. I thought, *Geez, this is wonderful*. Everybody who played on that team who applied in 1978 got in in 1980. Someone once asked me what I owed getting selected to? I said, 'Softball has been very, very good to me.'

---

[3] Jay Honeycutt joined NASA in 1966 and later served as director of the John F. Kennedy Space Center from 1995 to 1997.

MIKE MULLANE: [Prior to finding out I had been successful,] I turned on the news, and there was Anna Fisher. She was one of the six female astronauts who had been selected, and she was incandescent with joy that she had been chosen as an astronaut. And I knew that I had not been selected, because I thought I would have heard by then. I'll admit I was crushed, but I had a job to do. I was flying that day, so I couldn't dwell on it. Thank God, in the office somebody told me my wife had called, and Donna told me that Mr Abbey from NASA had called and wanted me to call him back. When I heard that, I still thought it would be, 'Thanks but no thanks. We appreciate you going to all that trouble. Sorry we couldn't select everybody.' You know, one of those calls to let you down easy. So, I called him with the mindset that I was going to be let down easy. He then asked me if I was still interested in coming to NASA to be an astronaut. I thought, *This doesn't sound like a rejection call.* I was absolutely overjoyed. I was already weightless – you didn't need to put me in space. I was beyond joy. The only other time I felt like that was when I got married to Donna and knew that I was going to be happy for the rest of my life.

When the 35 people out of the 8,000 who had applied were [formally introduced] to the world as having been selected in the first group of space shuttle astronauts, I rapidly learned that the white males were invisible to the press. Once our names were announced, the press swarmed the stage, but not to see Mike Mullane. They swarmed the stage to be around the six women and the three African American astronauts. I could have stripped my pants off and mooned the press, and nobody would have even taken a picture. I wasn't complaining, though. I didn't want to talk to anybody, because, to me, a camera's like an aircraft weapon pointed at you in Vietnam – dangerous.

I had never worked with civilians or women before, so I was suspicious of whether they were going to be able to really fulfil the role of being an astronaut. I had been in combat, and these people had done nothing but studied and written theses and been in laboratories and stuff like that. *There's no way that a civilian, I don't care who they are, male or female, is going to be able to equal that flying experience* was what was going on in my brain at that moment. Boy, was I wrong. Looking back on it, I can see the ridiculousness of thinking that I was better than them just because I'd had this flying experience. They were incredibly smart people, and I didn't realise until years later the history I was surrounded with. There was Sally Ride, the first American woman in space. I flew with Judy [Resnik], the second American woman in space. I was going to parties and palling around with Ron McNair and the other African American astronauts who were making history. And I feel privileged that I was around people like that. I was arrogant. That's the word. I was arrogant.

The 'Original Seven' were called that because they were the first NASA astronauts. That was the name of their class. We were the new guys, which turned into the TFNGs, or 'Thirty-five New Guys'. But 'FNG' has an obscene meaning in the military. Whenever you got to a squadron, you were called an 'FNG', a 'fucking new guy'. So, we were called the TFNGs by the public and the press, but, secretly, we knew we were the fucking new guys in the astronaut office.

ASTRONAUT CANDIDATES SELECTED JANUARY, 1978

 BLUFORD
 BRANDENSTEIN
 BUCHLI
 COATS
 COVEY
 CREIGHTON
 FABIAN

 FISHER
 GARDNER
 GIBSON
 GREGORY
 GRIGGS
 HART
 HAUCK

 HAWLEY
 HOFFMAN
 LUCID
 McBRIDE
 McNAIR
 MULLANE
 NAGEL

 NELSON
 ONIZUKA
 RESNIK
 RIDE
 SCOBEE
 SEDDON
 SHAW

 SHRIVER
 STEWART
 SULLIVAN
 THAGARD
 VAN HOFTEN
 WALKER
 WILLIAMS

THE NEW GENERATION

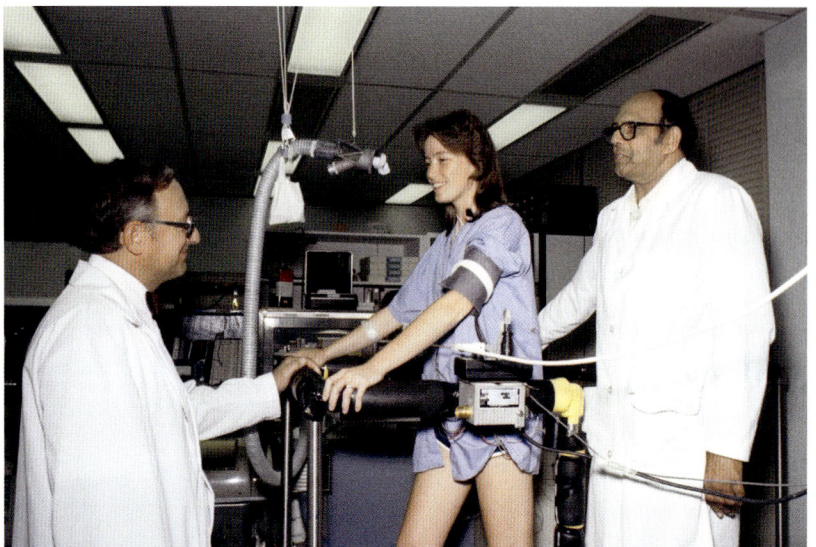

Below: Observed by Dr John Ziegelschmid (left), of the JSC's Medical Sciences Division, and environmental psychologist Herman S. Sharma, Anna exercises on a treadmill in the cardio-pulmonary laboratory at Johnson Space Center, 1977.

Bottom: Anna in training, 1980.

ANNA FISHER: I had such a sense of finally being where I belonged and where I wanted to be. We would have happy hours on Friday nights and did a lot of socialising together. Because I had studied so hard in high school and as an undergraduate and then in medical school, it was the first time in my life that I was able to really have fun. It had been a lot of hard work before that to get there, so it was really fun. Every so often I would have to stop and pinch myself and say, 'I'm just so lucky to be here.'

Of course, you're not a real astronaut until you've flown in space, so there was always a feeling that you were still pretending to be an astronaut until you had actually gone on a mission. It's sort of an unspoken thing that everybody wants to be first. That's the kind of people that are there. Astronauts will bet on everything. There was endless competition, but it was friendly competition. We knew it was going to be a couple of years before anyone in our group was assigned to a mission, but the competition started to increase when we had our first crew assignments. I think it became less fun for some people at that point if they thought they weren't going to be one of the early people to fly. We just wanted to be first, because that's how we were all wired.

MIKE MULLANE: NASA has the greatest training programme in the world. They are basically of the attitude that they're going to send you on a mission, and you are never going to be surprised by anything in the way of a malfunction or emergency. You will have seen it all and practised it all in your training before you ever go up there. That was their philosophy.

It was just such a thrill. Every morning it was a joy to wake up [and continue] training for a mission. I couldn't wait to get in the simulators. The garden of dreams was in full blossom. There was a feeling of, *This is going to happen.*

As he worked with the women astronauts in his class, particularly Judith Resnik, Mike Mullane found his attitudes transformed.

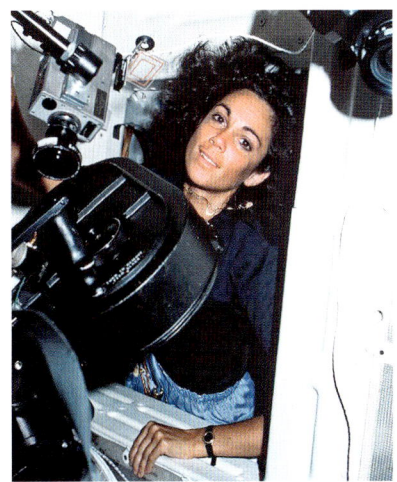

Below: Resnik is shown onboard the Space Shuttle Discovery during STS-41-D. Henry Hartsfield, the commander of the mission, described Resnik as 'an astronaut's astronaut... not satisfied with second best.'

MIKE MULLANE: Judy was a wonderful woman. I really changed my mind dramatically about the role of women ... in anything, frankly. On my first mission with Judy, we carried an IMAX camera. Henry Hartsfield was filming out of the back window when all of a sudden Judy screamed, 'Stop, Hank, stop.' Her hair had got caught in the belt drive and jammed the camera. Hank was going to report to the IMAX people down at mission control what had happened. Well, Judy looked at him and said, 'If you make that call, you're going to be minus some very important body parts.' She was going to cut his heart out with a spoon if he reported to mission control that her hair had got caught in the camera. And the reason was she knew – and she was absolutely right – that no matter what she did on the mission, no matter how capable she was, the press would have ignored all of it. All they would have talked about was her hair getting caught in the camera, and people would have been joking that this is why women can't fly in space.

The women were under an incredible microscope back then to not show the slightest difference between their performance and that of a man. Judy was on board with that – she wasn't going to make any mistakes, and she proved herself. She was all in on studying and working together as a crew and as a team, and I just had the greatest admiration for her, because she really made me a different person when it came to my attitudes toward women. My arrogance and my superiority complex about being a ten-year veteran of combat flying crumbled around her. I admired all of the women for the way they handled being the first six American women astronauts. Their performance in all regards opened my eyes to the fact that I wasn't this superior being just because I was a man. I look back on it, and I've got this sense of shame. How could I have ever had that attitude, you know? Anyway, I'm glad I changed. It was great. I really, really enjoyed flying with her.

Resnik flew with Mullane on mission STS-41-D on 30 August 1984, becoming the second American woman to fly in space after Sally Ride, who had been the first, flying in June 1983 on mission STS-7. Anna Fisher did not have long to wait for her turn, going into space on mission STS-51-A on 8 November 1984. By doing so, she was the first mother in space.

ANNA FISHER: When Sally Ride flew, I was eight months pregnant. George Abbey called me and Bill into his office and said, 'I'm thinking about assigning Anna to a flight.' I quickly said, 'I accept.' I didn't even give Bill a chance to answer, although there was no way that he would have said anything different.

I had Kristin at 9.30 in the morning on a Friday, and I was at the Monday morning meeting at eight o'clock. I wanted all my friends and colleagues and everyone to know that although I'd just had a baby, I was definitely going to be on the flight. I didn't really get any formal maternity leave, but for about four weeks my trainers put all my training on one or two days so that I could have some days off. After that, it was too busy, and we had to get back to a regular work week.

I definitely felt like I had more to prove as a woman and mother – that a woman could have a baby and keep going if she had to. I don't think it was the ideal way to do it, though, and I don't recommend it. However, if I'd said no, I might not have been assigned another flight for a long time. You just can't predict those things, and I had dreamed about it ever since I was 12. I felt a responsibility, both to NASA and to myself.

I know there were some people who were hostile and thought that what I was doing was wrong. How could I make a decision like that, to leave a 14-month-old baby? I've always said that it's the hardest thing I've ever done, but several of my male friends were in orbit when one of their children was born, and nobody criticised them for not being there. However, I didn't necessarily take offence. I understood the culture at the time, and I understood why some people might think [what I did was wrong], but I didn't think they really had any right to tell me what I should do. And it didn't affect what I was going to do, because I had made my decision and my commitment, and there was no way I was going to change that. Also, Bill was there if anything did happen to me.

Somewhere at the back of my mind I knew that it was possible that I might not be here for Kristin's next birthday. If those sorts of thoughts popped into my head, I'd quickly say to myself, 'What will happen will happen, and it'll be as it was meant to be.'

I never really thought that Kristin might grow up thinking I chose space over her. I guess I just assumed she would understand. Of course I wanted to be there with her. Occasionally I thought, *I've dreamed about this for so long, but is it really going to be worth the risk?*

Since I knew that she wouldn't remember me [if anything did happen], we took lots of videos. And we didn't just do birthdays or things like that. We videoed ordinary moments, like going on a bike ride or swimming in the pool, so she would have that if anything did happen to me.

Right: Anna hugs daughter Kristin on her return to Earth.

THE NEW GENERATION

Kristin Fisher became CNN's space correspondent as an adult, but when she was growing up, it seemed normal that both her parents were astronauts.

KRISTIN FISHER: I grew up five minutes away from the Johnson Space Center in Clear Lake, Texas, in a neighbourhood called Taylor Lake Village, and everybody was an astronaut or an engineer or affiliated with the space programme, so I grew up thinking it was the most normal thing in the world to have parents who were astronauts.

John Young lived a five-minute drive away. He was the commander of the very first Space Shuttle mission, and he landed on the Moon [as the commander of Apollo 16]. He was a legend. We would take our boat out to his house, dock at the end of his front yard, and then go in and have dinner with him and his wife Susy.

When you're a kid, you think that space is cool, and you think that your parents are cool. You're still kind of in thrall to them because they're your parents. But I don't care who your parents are or what they do, when you become a teenager, they become a lot less cool no matter what they do. So, to give you an example, we had to watch every launch if we were at home. It didn't matter the time – it could be two in the morning, and we had school the next day – my mom would wake my sister and me up every time to watch the launch. And if you've ever watched my mom watch a rocket launch, especially if there are humans on board, it is a scene. Ten, nine, eight

… she's getting more and more excited. Five, four, three, two, one. She has this ridiculous saying: 'Godspeed, go, go, go.' I remember being 13 or 14, watching this with my sister, and thinking, *Oh my god, Mom, get it together, you're so embarrassing.* But it was her friends on those shuttle missions, and she should absolutely have been cheering like that.

It wasn't until I went away to college in Boston that I realised how special and unique my childhood had been. It was during the first few weeks of my freshman year, and I decided to try psychedelics for the very first time – magic mushrooms, to be specific – and I had an epiphany: *Holy shit, my parents are astronauts.* It hit me in an instant.

It was midnight, and I was deep into my little magic-mushroom journey when I decided to call my dad: 'Dad, oh my god, you went to space. What was that like?' He was a kid in the 1960s, so I think he knew I was doing what a lot of freshmen in college do, and he kind of laughed and said, 'Kristin, go to bed, call me in the morning and we'll talk about it.'

That night completely changed how I viewed space and space flight and my parents. I felt so homesick. I really missed them and remember wanting to call and tell them, 'I'm so sorry. I didn't fully appreciate what badasses the two of you are.' The next morning, I called them and conveyed as much.

NASA ASTRONAUT GROUP 9

While Anna and Mike underwent their training, the astronauts of Group 9 were selected in 1980.

FRANKLIN CHANG-DÍAZ: I saw an announcement in the MIT newspaper that NASA was recruiting a new group of astronauts for the Space Shuttle programme. It was looking for young scientists in good physical condition who were American citizens. I had just become an American citizen, so the letter that I had received from NASA back in 1967 with the underlined sentence ... well, I'd just checked that requirement.

Within a month, I got a rejection letter. It was hard. It was a very terse letter: 'You didn't make the selection.' But I've never really succeeded [first time]. I've always had difficulties, and I've always failed. To me, it's a way of life. And I felt that this was just another one of those times.

I then almost forgot about the space programme again, because I was working to develop a fusion reactor that would harness the power of the sun on Earth and provide virtually unlimited power for humans, which I thought was extraordinary.

Two years later, in 1979, NASA issued a call for a second group of shuttle astronauts, and I pulled out my old application and waited. It was a few months later that I learned that I had made the cut from about 3,500 applicants down to the 120 interviewees.

When I got there [the Johnson Space Center], there were all these military folks – test pilots, marines and all of that. I think I was the only scientist. One of the military candidates was

Charlie Bolden. That was the first time we met. He just glowed. When the guy got out of the plane, it was like the whole room lit up. Everybody knew him, and he knew everybody. He was extremely gregarious, whereas I was very intimidated by all of these folks who were boasting about what aircraft they flew. I had to learn to connect to this new group of people.

Of that interview group, I think Charlie and I were the only ones who were selected. Finding out a few months later was a remarkable moment in my life – I remember every second. The first thing I did was pick up the phone and make a long-distance international call to Costa Rica to tell my family the news. My father answered, and my mother was right next to him. They owned a gas station at the time. I said, 'I've just been selected as an astronaut,' and my father started to cry. He couldn't talk, so he gave the phone to my mother, and she started screaming and jumping up and down and telling the pump operators and everybody who worked there, 'He made it.' I was elated. I was flying.

There was a whirlwind of attention, and in Costa Rica, I was famous. I instantly became a public figure. It's something that I wasn't prepared for, as it's one thing that I didn't really plan for in my very carefully planned life. I felt a sense of invincibility and that I was somehow protected from anything bad happening. We were still innocent. The bad thing hadn't happened yet.

Above: Jeffrey A. Hoffman and Franklin Chang-Díaz, wearing breathing apparatus masks during a pre-breathe session, on the forward flight deck of the Space Shuttle Atlantis during mission STS-46, 1992.

Left: Anna Fisher, training for a spacewalk in the pool at Johnson Space Center's neutral bouncy lab pool to simulate zero gravity.

Above: Franklin Chang-Díaz about to enter the pool.

CHARLES BOLDEN: I was at the United States Naval Test Pilot School, located at Naval Air Station (NAS) Patuxent River, Maryland, not very far out of DC. It was a Friday, and it was the beginning of our reunion weekend. We had got wind that there were three or four NASA T-38s coming in. Half the damn guys at the test-pilot school wanted to be astronauts, so everybody wanted to see who came in. That was when I saw this Black guy get out of the back seat of one of the T-38s. I remember thinking, *Wow!* I rushed out to meet him and introduce myself.

By that time, I knew quite a bit about Ron McNair from reading about him. I took him home for dinner to meet Jackie, my wife, and our kids, and then we got together several times over the course of the weekend. He was gracious when he came over for dinner and very generous with his time. There were no other Black people at the test-pilot school, plus we were both from South Carolina and our mothers were both teachers, so we had a lot of things in common.

I was mesmerised talking to him. As he left to go back to Houston, he asked me if I was going to apply for the space programme. I said, 'Not on your life.' He looked at me real strange and said, 'Why not?' I replied, 'They'd never pick me.' He paused for a moment and then said, 'That is the dumbest thing I ever heard. How do you know if you don't ask?' That hit me like a bullet. I'd forgotten that my mom had told me I could do anything. Why was I now doubting that? That's when I went home and told Jackie, 'I don't think I'll be picked, but I don't want to wake up ten years from now and say, "Man, what would have happened if I had applied?"' That's why I say Ron McNair is my idol and my role model, because he painfully reminded me that I had forgotten what my mom and dad had taught me growing up – that I could do anything I wanted to do. Before that, I never even gave any consideration to applying for the astronaut programme, even when NASA advertised in 1977.

So, I applied and ended up getting an interview and being selected in the second group of space shuttle astronauts. But that would have never happened had I not met Ron McNair.

I had no problems being a Black man in Texas until I had to go to a city called Pasadena. I was driving there when I saw a giant billboard with the image of a hooded man on a white horse rearing up – welcome to Pasadena, home of the Ku Klux Klan. This was July 1980. I couldn't believe it. I thought we were done with this! And although Houston was like an island in the state of Texas in terms of diversity and inclusion, it was still a southern town. There were no places you couldn't go, but there were places that were uncomfortable to visit.

When I got there, Ron was one of the first people who took me under his wing. He knew all of the nightclubs in Houston, and he and [his wife] Cheryl had a very active social life. Ron played the saxophone and was a very accomplished jazz musician. He was also a unique dresser and wore really contemporary clothes. Ron could get away with it. The rest of us couldn't.

Above: (from left to right) Charles Bolden, Gulon "Guy" Bluford, Ron McNair and Fred Gregory, with a shuttle model c. 1980s.

BILL FISHER: I was mowing my lawn on 30 May 1980, a hot and sweaty day, when Anna came out and said, 'Mr Abbey's on the phone for you.' Because Mr Abbey didn't call with rejections, [I knew that it was good news]. I have a bulkhead in my backyard, and I took a hammer and nail and hit the nail right into the middle of the bulkhead to remember that moment. It's still there. Then I went and took the phone call, and George said, 'How would you like to join us?'

I felt the same way I had when I'd received my letter of acceptance to medical school. I'd read that letter and broken into tears, and after I hung up with George, I did the same thing, because all the hard work was suddenly worth it, and there was tremendous relief and emotion and joy that you're going to get to do what you've wanted most to do all your life.

Alan Bean, the fourth man on the Moon, took care of our class, and one of us asked him, 'What did it feel like to walk on the Moon?' That's a wrong-stuff question. You never asked how it felt to do something. Bean gave a great answer: 'I look at a map and see Minneapolis, and I've been to Minneapolis, and I look at the Moon, and I've been to the Moon. It's just the same.' You don't give an emotional answer.

It was a funny kind of macho environment where you did your job. You were never afraid. You pushed ahead. You did what you were told. If you had to improvise, you improvised. But you didn't cry. Just do your job. Be competent. Do not display emotions that might have been interpreted as weak.

KRISTIN FISHER: I think the Right Stuff has changed over time. The Right Stuff for the Mercury astronauts was you've got to be a fighter pilot willing to fly by the seat of your pants, faster than the speed of sound, and take the biggest risks known to mankind. To have the Right Stuff in my parents' time was to be more of a team player. You were going to have six or seven crewmates on the Space Shuttle with you, and you had to play a role as you would in a family.

[My father] was really different [from the other astronauts]. He was very anti-authority, liked to be his own boss, did not do well in a bureaucracy. And, of course, NASA is a very big bureaucracy, which is what I think ultimately pushed him out of the programme. But he's an engineer and a medical doctor. He's also a student of history, and he writes. He's very romantic and a poet. What other astronaut has lived in a nudist colony in Hawaii during a summer in between medical school? But he knew how to have the Right Stuff, or at least how to hide the wrong stuff so that it appeared he had the Right Stuff for what it was in the 1970s and 1980s.

Right: Bill Fisher in spacesuit training.

A year behind schedule, the first shuttle mission finally launched on 12 April 1981. The success of STS-1 was an important milestone for the Group 8 and 9 astronauts, who were now one step closer to making their space dreams a reality.

CHARLES BOLDEN: There was nothing like the shuttle. It shook the Earth and seconds after you saw it lift off, you heard this rumble and roar and deafening sound, and then you got a pounding on your chest as the air vibrated from the pulses coming from the solid rocket boosters. It was unlike anything I'd ever seen or experienced. It was mind boggling to see something leave the planet. That is not normal. I think it is among the most fantastic, phenomenal things humanity has ever done.

[However,] I thought it was crazy that NASA declared the shuttle operational. That might have been good enough for the public, but we knew it wasn't the case. The shuttle was never going to be an operational vehicle. We were flying an experimental vehicle.

Right: Columbia sits on Launch Pad 39A before its maiden flight in April 1981.

Next page: An early morning scene at the Kennedy Space Center's Launch Complex, 1981.

52 THE NEW GENERATION

MIKE MULLANE: When the mission before ours landed, we were designated 'prime crew', meaning we were going to be the next people who were going to ride an American rocket off the planet. That was a powerful jolt to me: *This is for real*. I would wake up in the middle of the night with my heart going boom, boom, boom in this sudden overwhelming, powerful awareness that it was going to happen, that I was going to fly in space.

About three days before the mission, we got in our T-38 jets and flew to Kennedy Space Center.[4] The wives were out there to watch us arrive in Florida, and then they would come and give us a big hug. It was so exciting knowing that I was going to be riding the rocket in just a couple of days.

The day before a mission, the spouses would be brought out [to the beach house],[5] and we would sit around and have a meal, and then we would all drift off to the beach and say our goodbyes. And in the back of your mind, you know it might be the last goodbye, and they know it too. As always, I tried to use humour and jokes to camouflage the seriousness of the moment. Donna finally said, 'Just stop it and hold me.'

I told her that whatever happened, I was living a dream. If God had showed up at that point and said, 'There's a 50/50 chance you're not going to come back,' I would have taken that in a heartbeat. I was not going to back out. Now, I don't know [what I would have done] if Donna had said, 'If you do this, I'm going to divorce you.' Actually, I do know what the outcome would have been: I'd still have taken the ride.

There's a greater fear than dying for astronauts – the greatest fear is not taking the trip into space. I had to do it. The American public saw us as these heroes and heroines out there for God and country, putting our lives on the line to fly these vehicles. In truth, it's to satisfy a lifelong passion to fly in space. It is the most selfish act you can possibly imagine.

Left: The Astronaut Beach House, built in 1962 at Cape Canaveral, Florida.

Below: Members of the STS-41D flight crew are, from left to right, Michael L. Coats, Charles D. Walker, Steven A. Hawley, Judith A. Resnik, Richard M. Mullane, and Henry Hartsfield..

After you drive to the launch pad, you get out of the van and you're assaulted with this screeching noise, which is the engine purge.[6] And then you get into the elevator and take the ride to the 195-foot level. That's the cockpit level. It was June in Florida, damp, so there were mosquitos everywhere – we were slapping them off each other's backs. Eventually, we went into what they call the 'White Room', which was a box that was right up next to the side hatch. There we were dressed in the harnesses that would hold us in our seats. You then had to crawl on your hands and knees on a platform to get into the cockpit. This was all occurring about an hour and a half before launch. Then they close the side hatch, and at that point, there's not much to do until about ten minutes before lift-off. It's quiet. You don't hear anything except the soft whoosh of the cooling fans in the cockpit. At the five-minute point you can feel the vibrations of the hydraulic pumps.

Contrary to what people think, we do not hear 'nine, eight, seven, six'. Instead, we hear launch control announce, 'One minute ... thirty seconds ... ten seconds ... five seconds.' There's no way to calm yourself when you hear 'thirty seconds' then 'ten seconds'. My fear factor was off-the-scale high.

At T-minus six seconds the liquid engines start. You have one and a half million pounds of thrust coming to life, but it's being held on that pad, so there is a real growling noise with a high-frequency vibration in the cockpit. I remember thinking at that point, *Well, I'm going now.*

'Five seconds' ... then silence, except for a master alarm going off. Then all the engines shut off. There was probably five seconds of, *Oh my god, what's this all about?* But then we realised it was normal. The machine was designed to stop flying if it detected a problem. Then they mentioned fire [over the comms]. When you're sitting on four million pounds of propellant, believe me, the word fire registers big time in your brain.

---

[4] Mission STS-41-D was due to launch on 26 June 1984.

[5] A retreat at Kennedy Space Center where the astronauts could be alone with their families before a mission.

[6] The use of helium and nitrogen to remove moisture and air from the Space Shuttle Main Engines (SSMEs) prior to launch.

DONNA MULLANE: The morning of the launch, we were picked up by a bus to take us to the launch control building, where we watched a monitor. When it got to T-minus nine minutes, they said, 'It's time to go on the roof.' So we went upstairs to the roof of the building, and the shuttle was there, three miles away. It was sort of foggy.

Two astronauts [were assigned to] take care of us during the time we were at the Cape.[7] Ours were Dick Covey and Bryan O'Connor. We called them 'escorts into widowhood'. If something bad happened, they would take over.

So, we were on the roof, and the countdown started. I thought, *Oh boy, here we go*. Then, at [T-minus] four [seconds], it stopped. All I could see was smoke enveloping the shuttle. Then I heard this boom, boom, boom, boom coming towards me. That's what it was like – an explosion sound that arrived after I saw there was no more fire underneath. I thought, *It's exploded. That's why there's no fire underneath. It's exploded*. We were stunned. I thought, *Oh my god, are they gone? Is it over?* Even the astronaut escorts were confused. My prayers were going as fast as they could, and I was holding on to my kids. It was terrifying. It really was.

They finally got word to us that they had had an engine shut off and everybody was fine. I took a big sigh of relief. I don't even remember how Mike got to us. I can't remember any of that. It's a blur.

It was a while before he found out that he would still be on the mission, and it would go two months later.

Previous page: A flight readiness test firing (FRF) of Space Shuttle Discovery's main engines on the 2nd June 1984, ahead of its maiden flight, STS-41D.

Above: Bus taking astronauts' families to the launch viewing point.

Right: The Earth's limb at sunset, showing the thin veneer of the atmosphere backlit by the setting sun.

---

[7] Cape Canaveral in Florida, where the John F. Kennedy Space Center is located.

Mission STS-41-D eventually launched on 30 August 1984.

MIKE MULLANE: It was really just the same routine again, with all of the goodbyes at the beach house. Everything was the same. 'Thirty seconds … ten seconds.' We were watching the countdown clock, and when it hit zero, there was no doubt we were leaving the planet. At that point, there was this sense of release – there's no reason to worry any more.

At two minutes and twelve seconds, the boosters burn out and separate, bang, and you see a flume of fire go across the windscreen and then, like a light switch, dead quiet, no vibration. You're now above the thick part of the atmosphere, so it's a silent ride from then on. Crossing 50 miles is when you're officially an astronaut, so we all broke out into cheers [at that point].

The first thing I saw was a mosquito that had got aboard trying to fly in weightlessness, so I slapped it and got that killed. The last thing I wanted was to be up there with a mosquito. Then we all saw tiny bits of wire and screws and washers floating up. Things had fallen into nooks and crannies, and in weightlessness it immediately started floating around. It did cross my mind, I hope that's not out of something that's really critical.

I was just so incredibly happy. And everything is beautiful. No matter when or where you look, you're going to have your breath taken away. All the nights that I was up there, I rigged my sleeping restraint so I could float at the window and look out. I didn't get a good rest up there, but I thought I could rest when I got back to Earth. There were going to be very few days I was going to be in orbit and be able to see these sights.

As you go around the Earth, you have 45 minutes of dark and then approximately 45 minutes of light. When you're on the dark side of the Earth headed east, the atmosphere serves as a prism and splits the sun's light into its component colours. So, the first indication of sunrise is this eyelash-thin, dark indigo arc. And that is there for a couple of seconds. Then you see the lighter blue start filling in, and then you get the oranges and the reds. The only colour that isn't visible is green. It's this intense rainbow, 100 times more intense than any rainbow you ever saw on Earth, that just holds there on the horizon before the sun finally rises. And when that happens, it blasts those colours away. It was just wonderful.

Overleaf: Earth's sunrise is seen from space in this image captured by an astronaut on the International Space Station orbiting above the coast of Venezuela, 2022.

THE NEW GENERATION

DONNA MULLANE: The day that Mike did finally launch, the process was the same. They pick you up in a bus, they take you out, and you say your goodbyes again. Then we went up on the roof, and this time it launched. And it was so wonderful. It was just so exciting. Our son was pumping his fist, go, go, go. It takes your breath away, it really does. I thought, *Well, Mike, you're on top of there. Just come back. Just come back.* I earned every one of my grey hairs.

Opposite: Space Shuttle Discovery sits on Launch Pad 39A prior to the launch of its STS-41-D mission, 1984.

Anna Fisher's first – and only – space flight was on mission STS-51-A, which launched on 8 November 1984. It was the second mission since Mike Mullane's in a matter of just a little over two months, setting a record for the fastest turnaround of a reusable spacecraft in history.

ANNA FISHER: We went into quarantine on Halloween. It was Kristin's first real Halloween, so I told my commander that I was going for a jog, and I drove home and took her trick-or-treating. She was just a baby, but we went to two or three of the neighbours' houses. Technically, I broke quarantine, but at least I got to be with her for that.

The morning of the launch, because Bill was an astronaut [by then] too, he had all the passes, so I knew that he was going to be there. We all got ready, and as we were walking out, I suddenly realised that Kristin was there somewhere [with my mom], but I couldn't see her, and I couldn't see my mom. I saw Bill, and I waved, and my mom saw my eyes and realised that I wanted to see Kristin, so she got out of the van [she was in], and I got to see Kristin. For some reason, that was it – that was what I needed. I needed to see her, and now [I thought], *Whatever happens, happens*. I wanted to see her one last time before I went off into space. I felt at peace.

We knew that there was a good chance we were going to scrub that day because of the weather briefing, and sure enough, that's what happened. In those days, the astronauts paid for the buses for their guests. So, when they decided to scrub, my first thought in the cockpit was, *Oh my gosh, now we have to pay for another day of buses*. My second thought was, *What's going to happen tomorrow? Will I be able to see Kristin again?* But my mom did the same thing the next day, so it all worked out.

A lot of the risk is in those first two minutes [after lift-off]. Once you ignite the solid rocket boosters, you can't shut them off, so if something goes wrong, you really don't have a lot of options. Each milestone you hit, you feel a little better. 'Solids are gone.' Then you hear, 'Negative return,' which is at about four minutes. But the call I was listening for was, 'Press to MECO [main engine cut-off],' which means you can lose two engines and still make it to orbit. Once I heard the 'Press to MECO' call, I thought we were probably home free.

However, I also realised within 30 seconds that I was going to be one of the astronauts who has space adaptation syndrome [SAS], which is just a fancy term for symptoms that are like motion sickness. I remember thinking, *I'm not really feeling great. Why exactly was it I wanted to do this?* But it was a small price to pay.

The first glimpse of the Earth was so surreal – thinking back to when I was 12 years old, and that this was what I wanted to do, and now I'm actually here.

Getting to operate the [robotic] arm in orbit was so much fun. We had two satellites that we were taking up to orbit, and then on days five and seven, for the first time in history, we were going to try to bring two satellites back to Earth. They had been deployed during an earlier shuttle mission, but they were totally useless because they were in the wrong orbit. No one had ever done that before. These satellites were huge – each about the size of a small school bus. It was the beginning of us realising what we could accomplish in space, and I think it really helped make us confident when we were building the [International] Space Station that we could handle these really big pieces of hardware.

Of all the things that people had to say afterwards, the thing that meant the most to me was that Dr Kraft, the flight director who had basically invented mission control, said, other than Alan Shepard's flight and Apollo 11, this was the mission he was proudest of.

The ride back was like Mr Toad's Wild Ride at Disneyland. I was so happy to see Kristin, and my first words to Bill when I got back were, 'It was all worth it.' And I still feel that today. It all worked out. I took a gamble, and I won.

Below: Anna Fisher looks out of the aft windows of Space Shuttle Discovery's flight deck for a view into the cargo bay during STS-51A.

THE NEW GENERATION 67

Bill Fisher flew on mission STS-51-I, the 20th Space Shuttle flight, on 27 August 1985.

BILL FISHER: We had been selected in January for a mission to deploy three satellites. [It was a] vanilla mission. But then, in April, NASA launched Syncom IV-3, a giant satellite weighing 15,000 pounds. It was supposed to go into geosynchronous orbit, but the rocket didn't go off, so it stayed in low-Earth orbit. Peter Miller was the chairman of the board of Lloyd's of London, and they had insured the satellite. I'd gone out with Anna to visit Lloyd's one time, so I knew Peter. I decided to call him and finally got through. I said, 'Peter, we can fix this thing. You insured it for $85 million. You've got nothing to lose. We can fix it.' And I don't know how much influence I had, but they finally agreed to try it. If you salvage something at sea, you get 10 per cent of it, but they wouldn't agree to that.

From April onwards, once NASA gave us the go-ahead, it was 24-7 training to figure out how to repair it. They had a real satellite that hadn't been launched, so we trained on that.

Just to get to go to space was wonderful, but to do a space walk was the ultimate thing. I'll tell you what I did. When you're outside in the suit, the communications device has three positions: one is whatever you say goes to the whole world; another is whatever you say goes just to the Space Shuttle; and then there's an off position. For a moment, I turned mine to off and said out loud to myself, 'I'm thirty-nine years old, and this is as good as it's going to get.' Then I turned the switch back on. That [moment] was just for me.

There's a timeline [in space]: three in the afternoon, you're supposed to be doing this; five in the afternoon, you're supposed to be doing that. It's not like if you do something quickly, you can have a little spare time. The only spare time I really had was when we were doing [one of the two] space walks. I was standing on the edge of the shuttle, and they had a problem with the shuttle's arm. They said, 'Bill, you've got forty-five minutes.' The shuttle flies upside down, so Earth is set as your sky. I got to stand out there for 45 minutes and just watch. Not everybody gets that luxury of time.

These are almost clichés, but you don't see any country boundaries. You can't tell where Mexico begins and the US ends, so you get a feeling of unity. It is a planet that's inhabited by human beings and other animals, and it's just unbelievably beautiful, like a kaleidoscope constantly changing with the lights coming at you in different ways.

Below: Bill Fisher during extravehicular activity (EVA).

Above: Astronaut Bill Fisher stands in Space Shuttle Discovery's cargo bay during STS-51-I.

Opposite: Pictured on Discovery's middeck with one of the Extravehicular Mobility Unit (EMU) space suits, astronaut Bill Fisher.

Charles Bolden and Franklin Chang-Díaz both went into space for the first time on the same mission, STS-61-C, on 12 January 1986. It was the 24th Space Shuttle flight.

FRANKLIN CHANG-DÍAZ: Every time the Americans were planning to do something new – send somebody of colour, or send a woman into space – the Russians would do it first. It was always like that. And then, of course, it was time for the first Latin American to fly in space. I thought it was going to be me, but it was Arnaldo Tamayo Méndez, who flew a few years earlier, and the Mexican astronaut Rodolfo Neri Vela flew about a month before me. In fact, I was the one who strapped him into the shuttle! So, I was not even number two; I was number three. So, I said, 'OK, the Latino part is not in the game any more, but I have a little bit of Chinese [heritage], so maybe I'll be the first Chinese [person in space]!' Well, it turned out that the Chinese American astronaut Taylor Wang flew before me. I thought, *That's cool. I'm going to be the first naturalised American astronaut.* But they flew an Australian astronaut before me. I got the best thing of all, though, which is that I flew more flights in space than anyone else, and no one has taken that away from me yet.[8]

It was a surprise to me to finally go, because we had attempted to fly so many times and had failed. We had

come as close as 30 seconds before lift-off, and the clock goes to zero and nothing happens. You're in this live spaceship with half a million gallons of liquid hydrogen and liquid oxygen about to be lit up, and it's scary to think, *What if this goes up?* We were attempting to launch in very adverse conditions. At one point we boarded the shuttle in a lightning storm. Like I said, we were innocent. We just didn't really realise how bad it could be.

It was extremely emotional for me to be in space, the place that I had dreamed about being my entire life.

I felt very fortunate. You know the beauty of what you're about to see, but it is hard to really connect the theory with the experience. The theory doesn't live up to it. Everybody else is out there, billions of people, and here you are. You're the only ones in this environment right now – out of all the human race, you're the only ones.

Above: Franklin Chang-Díaz, STS-61-C mission specialist, while checking cargo in the Space Shuttle Columbia's payload bay, turns to smile at a fellow crew member, 1986.

Opposite: Franklin Chang-Díaz's seven mission patches for his record-equalling number of orbital missions.

---

[8] Chang-Díaz holds the joint record of seven orbital space missions with fellow NASA astronaut Jerry Ross.

Ronald McNair flew on mission STS-41-B on 3 February 1984, cementing his status as 'local hero' in his home state of South Carolina, after his mission returned to Earth. His brother Carl remembers this time.

CARL McNAIR: Ronald went up on stage and said, 'It was a smooth ride. Can't wait to do it again.' We were there in a crowd, screaming and yelling. They knew that the McNair family was there, because there were a couple of rows of Black people – you couldn't miss us.

That same day, they named the main boulevard through my hometown after him. It's something I never thought I would witness in my entire life – not the fact that he was an astronaut, but that there were as many white people as there were Black people celebrating their hometown hero. I never thought I would see that, considering where we started. That was amazing. The library is now called the Ronald E. McNair Life History Center. It's really beautiful. It's incredible. From slavery to space in four generations.

Opposite: Ron McNair at a procession in his homecoming parade.

Above: Ron McNair signing autographs at the homecoming parade.

THE NEW GENERATION 75

SS CHALLENGER

Below: Anna Fisher near the aft flight deck of Space Shuttle Discovery during mission STS-51-A.

Opposite: Anna Fisher in her spacesuit. The iconic image is part of a series of portraits taken by acclaimed photojournalist, John Bryson.

ANNA FISHER: I don't want to say that the compulsion to go into space is akin to a religion, but there is a similar kind of a dedication to something that you really believe in. And I understand that not everybody has this irrational dedication to wanting to leave our planet, but it's the one thing that I think really unites all the astronauts. There's this common passion that you could perhaps call irrational, but it's there.

I knew at the time the risk I was taking, and if I had seen the Challenger accident before I applied to be an astronaut, I'm pretty sure I would still have gone ahead. As years go by, losing friends changes things a little bit, but I would still do the same thing over again.

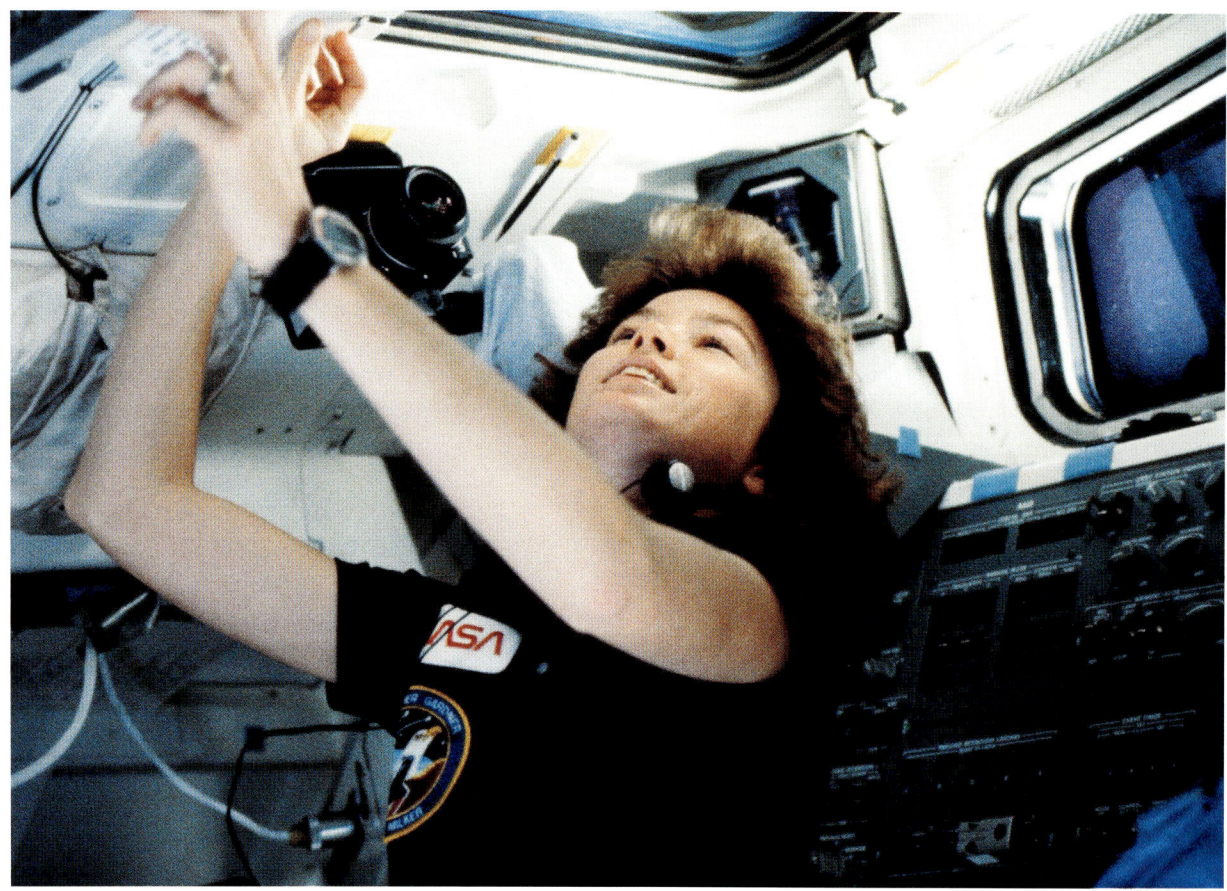

Below: Bill Fisher in EVA training.

Opposite: Anna Fisher in EVA suit training, with her daughter Kristin.

BILL FISHER: We wanted to go [to space], so if the shuttle was the way to get there, I felt fine with it, because it was the only option we had. I wanted to be a spaceman, and that was the only way to do it. I once heard a pilot say of flying in Vietnam that it was a bad war, but it was the only war we had. The shuttle might not have been the perfect vehicle, but it was the only vehicle we had.

The question was: is the risk worth the reward? And I think I speak for every astronaut when I say the answer was yes. At a pilots' meeting before the first shuttle flight, some engineers came and told us that they expected to lose one in 25 shuttle flights, or 4 per cent. The mortality of a Grand Prix race driver is less than 1 per cent.

Challenger was the 25th flight. We were constantly aware of it, but it wasn't a deterrent. It was certainly never a factor for me. It was worth the risk, without question.

KRISTIN FISHER: My mom has her own obsessions with space, but for my dad there's something about space that he has been attracted to from such a young age and pursued with such laser focus that he feels like his body, his spirit, his soul, does not belong on this Earth – it belongs somewhere else, somewhere out in space. I think for a lot of astronauts, it's a kind of spiritual calling. It's a compulsion, a pull from within that they feel. Certainly, from my father's perspective, it's something that he had no control over, that was inside of him from the time he was a small child. Is that obsession? Is that madness? It's a fine line.

78 THE NEW GENERATION

Below: Crew members of NASA's 41-D mission take a group shot to display their fun moments in space aboard the orbiter Discovery. They are (counter-clockwise from bottom centre) Henry Hartsfield Jr, crew commander; Michael Coats, pilot; Steven Hawley and Judith Resnik, both mission specialists; Charles Walker, payload specialist; and Mike Mullane, mission specialist.

MIKE MULLANE: The shuttle appeared to be exactly what everybody had envisioned: a reusable vehicle that would be easy to turn around for the next mission. So, the schedule started building quite rapidly, with crews being assigned and two shuttles being readied simultaneously for missions. [The Kennedy Space Center] looked like a futuristic space port, with these two monster machines a couple of miles apart on different pads.

There was huge scheduling pressure building. The shuttle was not going to make economic sense if it wasn't flying and flying often. And, in fact, the plan was 24 missions a year – a mission every two weeks. Schedule pressures pushed people over risk red lines and ultimately resulted in disaster.

I never talked privately with the other astronauts about safety concerns. I remember sitting in meetings where some issue was being discussed, and one of the first things out of my or the other astronauts' mouths would be, 'What's this going to do to the schedule?' In other words, how many weeks are going to slip to fix this problem? So, in a way, we were projecting the same sense of urgency: *We've got to get going!* We wanted to fly, and we wanted to fly often. That trumped everything else.

Below: Mike Mullane surrounded by cameras and Earth observation equipment on the flight deck of Space Shuttle Atlantis, 1988 In the frame are the Arriflex 16mm motion picture camera, a 70mm still camera, a 35mm still camera, a pair of glasses and a pair of binoculars.

FRANKLIN CHANG-DÍAZ: We had just landed Columbia on 18 January [1986]. They were in a hurry to land us so that they could launch Challenger, which was on the pad, ready to go. I felt that we were going a little too fast, but I guess I just didn't feel confident enough to speak up. I was a rookie. I had just done my first flight. There were people who were far, far more knowledgeable [than me], not just in operations, but in flight. There were astronauts who had flown, and I just didn't feel that it was my place to make any dissenting comments.

So, we landed in California [on the Shuttle's back-up landing strip], and that was the end of our mission. And then we headed back to the Johnson Space Center for our debriefs. We were still in the middle of debriefs ten days later, on 28 January, when we took a break to go and watch the launch of Challenger.

Right: The Space Shuttle Challenger on the crawler transporter, making its way to launch complex 39B.

On 28 January 1986, disaster struck for the first time in the Space Shuttle programme when Challenger exploded 73 seconds after lift-off, killing all seven crew members on board.

Left: Space Shuttle Challenger's frozen launch pad at the Kennedy Space Center, 1986.

Below: The crew of the Space Shuttle Challenger STS-5: (back row left to right) Ellison Onizuka, Charon Christa McAuliffe, Greg Jarvis, Judith Resnik, (front row left to right) Mike Smith, Dick Scobee, Ron McNair.

CHARLES BOLDEN: Challenger was scheduled to launch on 27 January,[9] but it was so cold they cancelled that day. The crew came back out the next morning, and it was equally cold, but the weather looked clear – it was just freezing cold. It was an environment in which we had never operated before.

BILL FISHER: It was freezing the day Challenger launched. There's no way it should have launched, but people from Washington were there, and it had already scrubbed twice, so there was tremendous pressure to launch. When the seals were cold, they lost their elasticity and were no longer able to do what they were designed to do. It was no surprise. In fact, when Challenger blew up, they knew exactly what had happened. But there was tremendous pressure, and the programme was being questioned.

CHARLES BOLDEN: Dick Scobee was the commander, and his pilot was Mike Smith. The mission specialists were Judy Resnik, Greg Jarvis, Ellison Onizuka and Ron McNair, and the final crew member was Christa McAuliffe, who was a teacher out of New England. It was an incredible crew.

My first thought was, *Please find a way to make this come out right.* I was hoping against hope that some miracle would take place. It was personal to almost everybody in the office.

---

[9] This was the second scheduled launch, after the planned launch on 25 January was scrubbed because of bad weather.

Below: Mike Mullane and Judith in training, 1984.

Next page: About 3 seconds into launch and 70 seconds before the Challenger exploded.

FRANKLIN CHANG-DÍAZ: I was just as mesmerised, as petrified, as horrified as everybody else was. Some people embraced and comforted each other and others walked out into the hallway like zombies. It was like it was the end of the world. Several members of the press were there, and they began to put cameras right in front of us. We were in no mood to talk to anyone. Eventually, NASA security was able to gather some control and keep everybody safe.

I had called Mike Smith, one of the crew members, the day before to wish him a good flight. That was the last time I talked to him. And from that moment, I never, ever called anybody over the phone to wish them a happy flight. I'm not a superstitious person, but I just couldn't do it. I was traumatised, perhaps. That's what it did to us. You got scars, and those scars never, ever stop hurting. They always hurt. We lost our innocence, and we bear those scars forever. You always go back and say, 'Could I have done something?' You grow old with the pain of those scars, and they remind you that you have to pay attention. That if there's something you can say, you should say it.

ANNA FISHER: I had already been assigned to my second flight, and we were about six weeks from our launch date. I was in the simulator, training, that day. We decided to freeze the 'sim' and go down to the conference room, where we normally did our debriefings, because there was a TV there. We heard the 'go at throttle up' call, and shortly after that, we saw what happened. We really didn't say very much, because all of us understood at that moment that there was no chance that anyone was going to survive.

I really wanted to see Kristin, so I got in my car, drove over to her preschool and picked her up. I remember telling her that there had been an accident and that I just wanted to be with her. As we were driving home, there was a contrail in the sky, and Kristin asked me if that was the shuttle. Suddenly, the things that we had all worried but not really talked about were real.

MIKE MULLANE: As soon as I saw the explosion, I knew the crew was gone. Four of them were in my astronaut class: Dick Scobee, Judy Resnik, Ron McNair and Ellison Onizuka. Four TFNGs killed on Challenger. They were my friends. I flew with Judy, and now she's gone, and Dick, Ellison and Ron are gone. But it could have been any of us. It was just the luck of the draw how the flights played out. It does weigh on you, though. Such a loss. I miss them all.

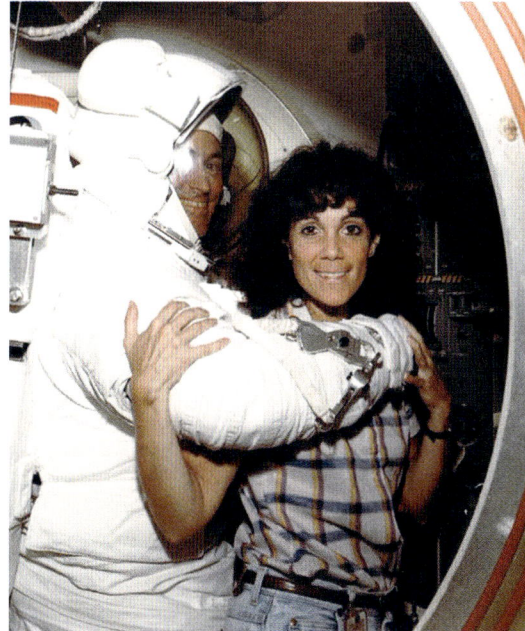

THE NEW GENERATION 85

Below: A trail of smoke leads up into the sky and then ends where the Challenger exploded 73 seconds after lift-off, 1986.

Bottom: Spectators in horror, shock and sadness after witnessing the Challenger explosion.

CARL MCNAIR: Ron came to my apartment in Atlanta, and he had this VHS cassette tape of the footage that he'd taken in space.[10] He was only interested in one specific part of the tape. He was fascinated with the launch itself, and he played the video up until the shuttle entered orbit. Then he'd rewind it and start over again. He kept turning the volume up and blew out one of my speakers. He wanted to feel the same kind of power and roar and sensation that he felt when it was lifting off the launch pad. The irony of it was that the scene on the video that he wanted to keep playing over and over again was the same point where he lost his life on the Space Shuttle.

One of the worst parts was that every day we saw them die over and over and over again. It was all over every news station. You were never, ever able to get away from it. I never knew where I might be or what I might be doing only to look up at a screen and see my brother dying. I would go places only to realise that the world was still going on, and people were still laughing and smiling. I thought, *Don't they realise what has happened to me and my family?*

BILL FISHER: I was flying out of Houston to go make a speech at some place in Daytona Beach. We were halfway there when the pilot announced, 'Ladies and gentlemen, I have some bad news for you' – [not the best thing to hear on an aeroplane] – and then he announced that Challenger had blown up. So, I didn't see it live; I saw it later on TV.

Someone said that you're not dead until the people who knew you are dead too, because your personality, your traits, your sense of humour, live on in the memories of others. All the people on Challenger, they're still alive in my heart and my head. I know what jokes they told. I know who they were. I know their families. Their personalities are still alive in me and the other people that knew them.

---

[10] Ron McNair was a crew member on two Space Shuttle Challenger flights: STS-41B in Ferbuary 1984 and STS-51L in January 1986, when he was killed in the first ever Shuttle accident.

Above: The Challenger explodes 73 seconds after lift-off, when one of the solid rocket boosters broke free.

CHARLES BOLDEN: Our boss, Mr Abbey, called me and said, 'Go out to Ellington [Field Joint Reserve Base in Texas] and get on a T-38 and head to the Cape. Take some clothes, because you're going to be the family escort for Ron's family. I wasn't quite ready to help a mother with two infants deal with her grief, so I had to learn real quick. I was grieving too, and still am. I don't know that you ever stop grieving for somebody who's incredibly close to you.

Left and right: Charles Bolden with Reginald McNair at Ron's memorial, 1986.

CARL McNAIR: Charlie meant the world. He and Ron were best of friends, and Charlie stayed by our side as often as he could, particularly with Cheryl and the kids. He did an exceptional job making sure that we had what we needed. He was there for us. If we had a question or needed something, we called Charlie, and Charlie tried to find a way to get it.

DONNA MULLANE: All I could think of was the families and how they were feeling and what they were going through. It was horrific. Judy [Resnik] was on there, and it broke my heart. It really broke my heart.

We went off to see the families. Outside of every home, there were four or five TV trucks in front of their lawns, perched waiting to see a family member come to the door or an astronaut arriving. It felt like a violation. Just leave them alone. Let them be in peace. It's hard enough that they've lost their family member. Just go away.

CHARLES BOLDEN: I went through this thought process of, *What do I want to do? Am I going to stay with this programme?* And it took me about a nanosecond to decide that Ron would kill me if I walked away now. So I said, 'I'm not going anywhere. I'm here for the duration.' I did feel like I owed it to Ron. He had been the person responsible for me being there, and I didn't want to walk away just because something bad had happened.

The Challenger disaster left NASA devastated and demoralised. *Where do we go from here? What are we going to do next?* We didn't have any clue as to how we were going to pick things up, because we knew that if we delayed too long, the interest in resuming shuttle flights would wane. Everybody would lose interest, and we'd have a hard time getting started again. We didn't want that to happen. And it took two, almost three, years, of no flying, and people wondering where in the world [the programme was] going.

THE NEW GENERATION 91

Below and opposite: The Challenger before the accident.

BILL FISHER: Chuck Yeager[11] said there was no more significance to this shuttle loss than there was to the loss of any test vehicle. He also said, 'We are going to lose people. That is the sacrifice you make. And our job is to fix the problem and get flying again.' People thought that was very insensitive, but I agreed with him.

In the movie, Rocky gets knocked down but that's not the end of the movie. No, he gets up and fights back. Just because you lose something, especially when you know it's a risky endeavour, doesn't mean you give up. You fix the problem and then, having learned from that, you do it again. Maybe later you'll lose another one, but every time you do it, you get better and better with the experience you've gained from your losses. We finally got flying again, but it was like Rocky got knocked down and didn't get up for two and a half years.

Challenger didn't put me off flying, because I knew we were going to fix the problem. I just thought we'd fly a lot sooner. As it was, we lost two shuttles in 135 flights – that's somewhere between 1 and 2 per cent, so it was better than one in 25.

---

[11] The legendary test pilot who became the first man to break the sound barrier in 1947, and featured in *The Right Stuff*.

THE AFTERMATH

The first flight after Challenger, on 29 September 1988, was mission STS-26, just over two and a half years later.

CARL McNAIR: Inside I was thinking, *Keep going and get past two minutes, as I think you're OK after two minutes.* It was overwhelming. It was very fulfilling. It was a hodgepodge of emotions to see that a crew was flying again.

I, like everybody else in the family, didn't want to see the shuttle programme discontinued just because of one accident. Not one of those astronauts would have wanted that. What my brother died doing is going to continue. It's not going to stop. That's what exploration is all about.

ANNA FISHER: When the Challenger accident happened, we knew that it would be a couple of years before we flew again. In the meantime, Bill and I decided we wanted to have our second child, so I needed to reassess things. This time, I decided I wanted to take some leave, which I had not done with Kristin. I wound up taking a seven-year leave of absence. And there were a lot of reasons why I did that, but one reason was that I realised being there for the two of them when they were young was really important. However, I still wanted to fly again.

The reason I wanted to do what I did was to be one of the early explorers who took humans out into the universe. Just to be a part of those first steps was worth the risk.

I think we [the TFNGs] were able to change that perception of what an astronaut could be and what they could look like. And the reason I think that's important is that when you look at our amazing planet from space, you no longer see the borders. You no longer see all the differences. It might sound corny or trite, but you really realise that we're all, in a way, astronauts on this planet, because when you're on it, it feels really big, but when you're up in space, it doesn't seem that big. And the more people who see that and realise that this is it, this is our home and we should take care of it and each other, the better.

I had an amazing experience, and I'm grateful for it, but I'm also grateful that I'm here to enjoy this time too, because 14 of my friends are not.

Opposite: This composite image of southern Africa and the surrounding oceans was captured over six orbits of the NASA/NOAA Suomi National Polar-orbiting Partnership spacecraft on April 9, 2015.

KRISTIN FISHER: [Mom] was assigned to a second space flight and then Challenger happened, and she was at an age where if she didn't have another child, she was never going to. So she got pregnant and decided to take a leave of absence from work when my sister was a baby. I also wonder if risk played a part. I haven't point-blank asked her, but we have had a hypothetical conversation about whether I would go into space right now. And I told her I'd honestly have to think about it, because I've got two really young kids. Her response was, 'I think this vehicle's pretty safe, and I'm sure you'd be fine.' So I think she knows exactly what was at stake when she flew. And I think she'd make the same decision all over again.

MIKE MULLANE: There were a lot of factors at work in me deciding to retire after my third mission. One was the effect the missions were having on Donna. She's not a woman who was built for high-stress environments like the shuttle programme. There was a part of me that thought, *It's ripping her up.* I guess I was getting older too and realised that there were other things at risk here – my family, my children – so it all came together. It just seemed like the right moment.

I remember all the nights I spent in space. The last two missions both went over North America, and I was able to see Albuquerque, where the dream was born. I could see the desert where I tested my homemade rockets and where as a kid I had watched Sputnik. It was like watching the movie of your life. It was powerfully moving to see where I started, and here I was now as an astronaut.

[On the last night in space,] I recollected all that had happened, from the beginning to the end. I obviously thought about Challenger, and all of the people I had met, and, sadly, the people who had died. At that moment, I had resolved in my own mind that leaving after this flight was the right thing to do. There were other horizons to look over. I now find plenty of time to go visit the kids and the grandkids. Donna and I do that a lot, and I love doing that.

DONNA MULLANE: Mike flew three times, but I went to the roof of that launch control centre nine times with him being strapped in. So, I had to say goodbye to him nine times. That takes a toll.

He had already told me that this was going to be his last mission. He had been an astronaut escort for a mission, and I think he realised the stress factor on a family. I don't know if that helped him make the decision that he wasn't going to do it any more. I think if he'd stayed on, he would have flown again. He could have possibly had his dream of a space walk, but he thought it was time to move on. I didn't say, 'Oh no, stay on.' I didn't say that.

I will never, ever say it wasn't a highlight in our lives, because it was. It was a wonderful experience. But would I want to do it again? No. I was glad when he said it was over.

Right: Space Shuttle Atlantis is seen as it launches from pad 39A on Friday, 8 July 2011, at NASA's Kennedy Space Center on the final flight of the space shuttle.

PART TWO: LIVING IN SPACE

'We did it for the whole world. We did it for the benefit of all humanity.'

ANATOLY ARTSEBARSKY,
Cosmonaut

INTRODUCTION

1986 was a strange yet undeniably pivotal year in the annals of human space exploration. It began with tragedy: the Challenger, one of America's new breed of advanced space shuttles, exploded just 73 seconds after lift-off, killing all seven of its crew. While NASA and the country were still reeling from the accident, the Soviet Union defiantly launched the Mir space station. This huge microgravity research laboratory, designed with the explicit goal of advancing technologies for permanent human occupation of space, began orbiting Earth every 90 minutes. Almost continuously inhabited by cosmonauts recruited from across the USSR, Mir quickly became the pride of the Soviet Union and the most sophisticated space station the world had yet seen.

It was a return to Soviet prominence not seen since the dawn of the Space Race (think this is fine but this is the kind of statement we would want checked by an expert). Indeed, it was the USSR that launched the first man-made satellite, Sputnik, back in 1957, igniting the fierce competition with America. What followed was an exceptional array of firsts, each demonstrating the Soviet Union's scientific prowess: two years after Sputnik, they launched the first rocket to leave Earth's gravity and orbit the sun; a few months after that, Luna 2 became the first man-made object to impact the surface of the Moon; in 1960, Sputnik 5 returned the first animals alive from space; less than a year later, Yuri Gagarin became the first human to orbit the Earth; two years after that, Valentina Tereshkova became the first woman in space; and the first ever spacewalk was performed by Alexei Leonov in 1965. The progress of the Soviet space programme was profound. While America would impressively fight back to become the first nation to land a man on the Moon in 1969, the Soviet Union did not relent, and in 1971, launched Salyut 1, the world's first space station. Fifteen years later, Mir launched, becoming the largest, most advanced space station ever to be built. At this stage, nobody knew more about off-Earth living than the Soviet Union.

On Earth, however, the situation within the USSR was not as triumphant. Economic uncertainty coupled with societal unrest was threatening to break up the Soviet Union. Inflation was soaring and many citizens were living in poverty. Even the space programme was feeling the pinch as the reality of economic collapse began to hit. One historical source of income for the Soviet space agency had been the sale of space seats, typically to potential cosmonauts from ideologically aligned countries. Known as the Interkosmos Programme, it was now expanding to accepting

Previous page: The Space Shuttle Atlantis docked at Russia's Mir space station, 1995.

offers of hard cash from any nation, regardless of its socialist sympathies. It was for this reason that a forty-something, chain-smoking television reporter became the first Japanese national in space in 1990. At the same time, he also became the world's first journalist to report from outer space. The experience would leave an indelible impression on Toyohiro Akiyama, perhaps not quite as you might imagine, as I discovered when I eventually met him – a virtual recluse now, living in rural Japan.

When the Soviet Union eventually fell apart, the dismantling of the old order began, with statues and flags of the old regime being torn down across the cities and towns of the newly independent countries. There was one flag of the old Soviet Union, however, that was too remote for even the most ardent of reformers to reach. As Commander Anatoly Artsebarsky was looking down from Mir on his beloved Soviet Union imploding below him, he wanted to make one final, patriotic statement. On the last spacewalk of his mission, he unfurled the hammer and sickle flag of the Soviet Union, which he had smuggled to the space station, by hiding it in his spacesuit. Without seeking permission from his superiors, he attached the flag to the exterior of Mir. There it hung, the red flag of the former USSR, orbiting Earth, attached to a giant Soviet space station.

Recounting that memory today, Anatoly is moved to tears. For him, Mir will always be the greatest achievement of the Soviet Union.

America's concern, however, looking on as the USSR collapsed into chaos, was the fate of its nuclear weapons. The Cold War might not have been ideal, but at least it was stable – both countries had weapons of mass destruction, both knew how to make them, and both maintained a space programme that kept their rocket scientists fully employed and focused on space. The thinking now, with the collapse of the Soviet economy, was grim: a brain drain of talented rocket scientists might leave Russia and wind up in countries even more ideologically opposed to the West. America's worry was that once in the hands of these regimes, their attention would shift away from space and onto missiles – rockets that went sideways instead of upwards. This was not baseless paranoia; plenty of nations would pay handsomely for the knowledge these Soviet rocket scientists held in their heads. For the Americans and the West, this was a huge threat to world peace.

A plan was needed and a plan was hatched. President Bill Clinton outlined a bold new strategy to keep the Russian space agency afloat. The US would fold their faltering American-only space station Freedom, and

Opposite: Mir space station, taken from space shuttle Atlantis.

begin a joint venture with Russia to build a new International Space Station and foster cooperation between the two countries.

To kick things off, the two agencies first needed to learn how to work together, and share valuable knowledge on their respective areas of expertise. A programme called Shuttle-Mir by the Americans, and Mir- Shuttle by the Russians, would see NASA's space shuttles deliver American astronauts to Mir to gain valuable knowledge on a practice the former Soviets were beginning to master: living in space for months on end. In return, the United States would give the Russian space agency a lifeline of desperately needed cash that would not only keep Mir safely in orbit, but also keep Russian scientists in full- time employment. Conceived as the first phase in achieving the long-term goal of building the International Space Station, this blended space venture promised to answer both nation's prayers. The only stumbling block was that these two former enemies would have to drop their long-held distrust of each other even to consider collaborating in space.

When astronaut Charles Bolden was told of the plan to send him to Russia to work with cosmonauts, his reply was unequivocal: 'I am not working with no damn Russian!' For an ex-fighter pilot who had fought Communists in Vietnam, the idea of cooperating with the Russians was a stretch too far. It was a reaction common among many of the astronauts, but not all. Michael Foale, a physicist and NASA astronaut, agreed to visit Russia for what he thought was just a week-long trip, intended to strengthen ties between the two space agencies. When he arrived, he discovered that the Russians were under the excited impression that he was moving there permanently and would soon be bringing his young family out with him. Known affectionately amongst friends as the perfect diplomat, Michael found himself agreeing to this, much to the delight of his hosts, and wondering how he was going to break the news to his wife Rhonda when he got home.

In 1993, when the Shuttle-Mir programme was announced, nobody knew if this experiment in collaboration between Russia and America would work. Never before had two former enemies put aside their political differences, to work together, in the joint pursuit of exploration, discovery and the furthering of human knowledge. Even more remarkable, was that this was all happening 200 miles above the Earth's surface, in an aging Soviet-era space station, orbiting the world every 90 minutes and travelling at speeds of approximately 17,000 mph.

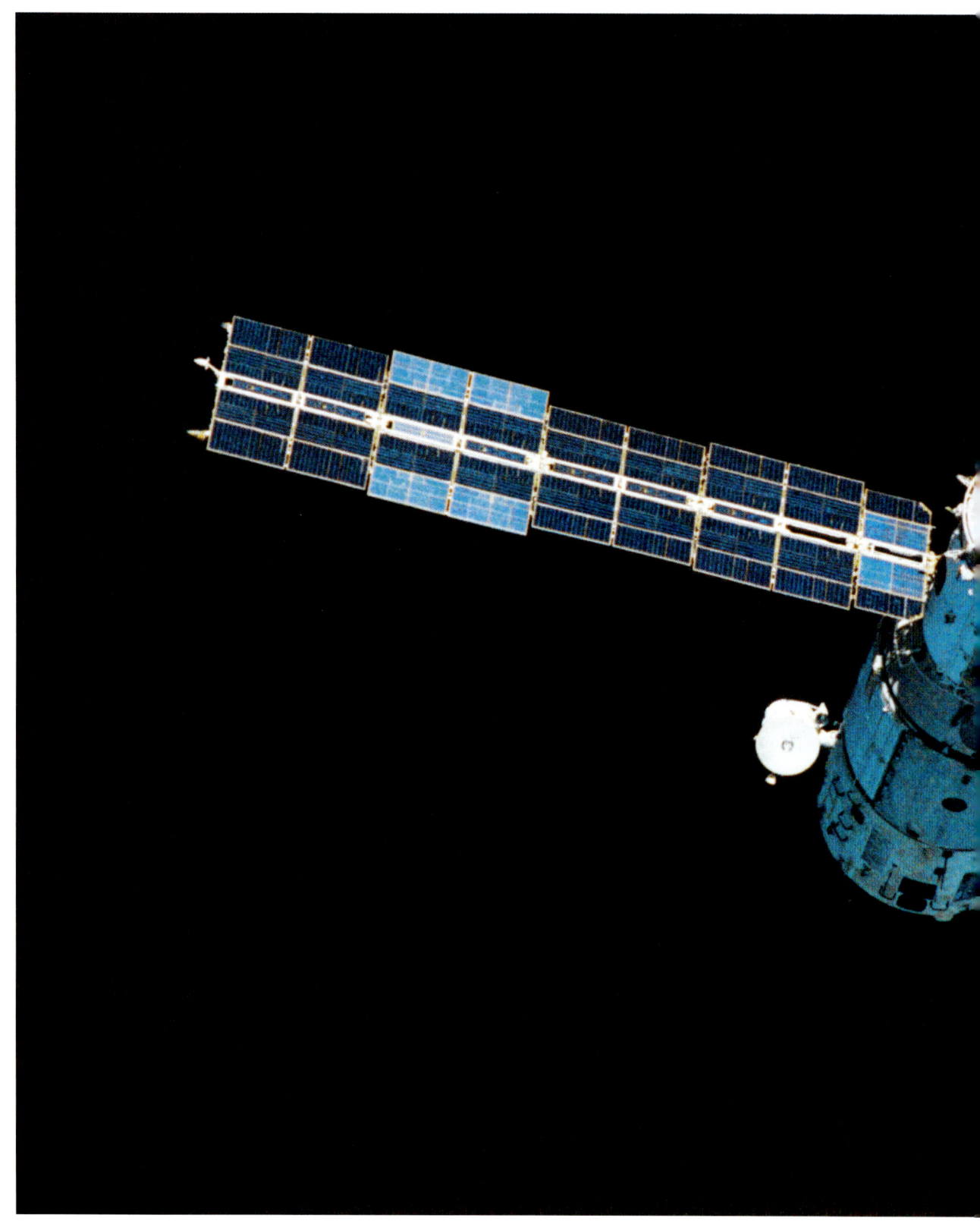

Above: Mir space station, taken from Space Shuttle Atlantis.

LIFE ON MIR

Anatoly Artsebarsky was the commander of Soyuz TM-12, which launched on a mission to Mir on 18 May 1991.

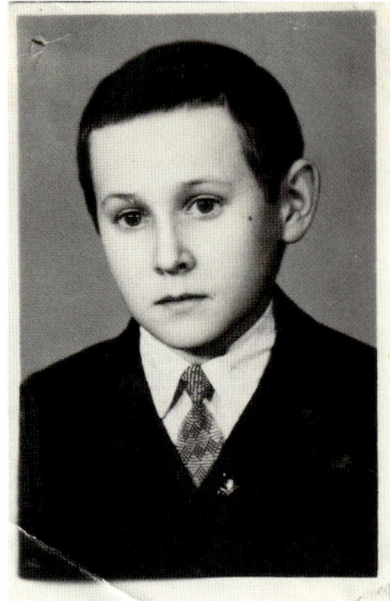

ANATOLY ARTSEBARSKY: I was four and a half years old when Gagarin went into space. My father took my brother and me outside that evening and showed us the starry night sky. We looked up very attentively, and it seemed to me as if a bright red rocket was flying across the sky with 'USSR' written on the side. From that moment on, I decided I wanted to be a cosmonaut. And that's exactly how it turned out.

I knew all the latest news about the American astronauts because I had a personal interest and listened to *Voice of America* from Washington on shortwave radio. This was not something that was welcomed in our village. Our local public prosecutor lived in the house next door, and he would walk by the window and listen in. He then informed on me in school. He said, 'Do you realise what one of your kids is listening to?' I had to be very careful, but I kept on tuning in.

When Kennedy announced that in the next ten years the Americans would land on the Moon, it's possible that no one believed it, but personally I did and even now, every 20 July, I celebrate the American Moon landings. It is important for me and has been ever since my childhood. The fact that the Americans were able to do that deserves only the greatest honour and respect. We also wanted to do that, but we didn't manage it.

The Americans launched their own station called Skylab,[1] but only three crews stayed there, and then it was deorbited. They were unable to create a multifunctional, long-running station like Salyut 7, say, or Mir. Even when it came to the ISS [International Space Station], they took a lot of inspiration from Mir.

Salyut was a series of space stations: Salyut 4, Salyut 5, and so on. They could house people who worked there and waited for another craft to dock. It was only the Mir station that had many docking modules – four spacecraft could dock there simultaneously.

The word 'mir' is very important to me – it means the 'world' or 'peace'. We did it for the whole world. The Soviet Union would fly anyone up there: there were Indians, French, Italians. We did it for the benefit of all humanity.

---

[1] NASA launched Skylab in May 1973, and it stayed in orbit until July 1979 when it decayed and disintegrated into the atmosphere.

Artsebarsky spent 144 days in space, and completed six space walks. As well as Flight Engineer Sergei Krikalev, Helen Sharman also travelled on Soyuz TM-12 and in doing so, became the first British person to fly to space. Her flight was part of Project Juno, a privately funded space programme.

ANATOLY ARTSEBARSKY: Under Project Juno, there was Helen and three other candidates. Helen was very pleasant, and we wanted her to be selected. We knew that she worked as a chemist for the Mars chocolate company and that was it; we didn't know anything else.

Everything that concerns Helen was a first. She was the first British woman in space. And she was the first female cosmonaut I had ever worked with. We were given no special instructions from anyone, we didn't know what to do and how. We just accepted the task and implemented it. There was Sergei Krikalev, me and Helen. She was well prepared, and she spoke excellent Russian too, which was important. That made it easy to chat and prepare. She never asked for help, she managed everything perfectly well on her own. I was very impressed.

If he ever hears this, I hope he forgives me for blowing his cover, but a colleague of mine was truly in love with Helen. I mean, for real. And, as I was the mission commander, he asked me to introduce him, but I didn't, because Leonov [2] had showed me his fist, as if to say, 'Don't let Helen come to any harm. Look after her!' A joke, of course.

Artsebarsky and Sharman spent seven days working together on board Mir before Sharman's departure on 26 May 1991.

When the time came for us to part, Helen came over to me and said, 'I want to stay here. I will never return here. Think of something so I can stay!' I thought, *'But how can I do that?'*. She left, with Afanasyev and Manarov. She had given us some music, including Tanita Tikaram's 'Twist in My Sobriety', which became my favourite song. I remember she had tears in her eyes as she said goodbye. I watched her ship as it departed. At first it was clearly visible and then it decreased in size until it was a small dot. And that was when we listened to the music.

The first time I was invited to come to Britain, 25 years after the mission, I arrived at the airport in London, and saw pictures of Tim Peake 'the first British astronaut' all over the place. Why Tim Peake? Our Helen was the first! Helen was a proper research cosmonaut. She completed her mission and her accomplishments should be more widely known. Tim Peake is a decent lad but he was not the first. And that's that.

Above: (from left to right) Anatoly Artsebarsky, Helen Sharman and Sergei Krikalev.

Opposite: Tense smiles from would-be astronauts at the Science Museum, London, as the countdown begins to the naming of the successful candidates for the Juno Space Mission.

---

[2] Alexei Leonov was the deputy director of the cosmonaut training centre and as a cosmonaut himself, had completed the world's first spacewalk on 18 March 1965.

Another visitor to Mir from outside the Soviet Union was Toyohiro Akiyama, a Japanese journalist who was not obvious cosmonaut material. After joining the Tokyo Broadcasting Service (TBS) in 1966, Akiyama worked in broadcast news, going on to become TBS's chief correspondent in Washington in 1984. He returned to Japan in 1988.

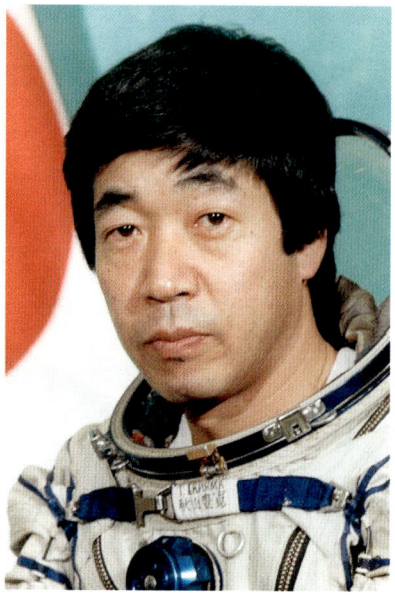

TOYOHIRO AKIYAMA: Later that year, I heard that a proposal to carry a Japanese person from TBS on board a Soyuz spacecraft to the Mir space station had been agreed by the Russians. I'm convinced that the Soviet Union wanted to use space to build stronger ties with various countries around the world, but it was also running out of money. I don't think they had decided on a fixed price for a ticket to Mir, but they could probably pay the annual salaries of a few thousand people working in the space programme if they sold seats to international guests a couple of times a year.

If Japan was going to send a journalist to space, I thought broadcasting everything live was the only way to go. It was ground-breaking; I thought I could give viewers at home a real sense of being in space. There was no way I would let a chance like this go. I would apply to go to Mir and do my best to be chosen. I was forty-seven years old, so I didn't mind dying – to die trying to broadcast live from space would be a noble way to go!

At first, people thought it was a joke that I wanted to put myself forward. I wasn't far off retirement age and there was a common preconception that candidates should be young. That made me furious. I felt like I was in my prime.

The Japanese medical team selected almost a hundred people, and it was narrowed down to twenty-one. Then when they were choosing the seven finalists, I was dropped. Someone came to tell me, 'You have traces of an ulcer in your stomach.' I thought, 'That's bullshit! A guy in his late forties, working for a TV station that doesn't have a stomach ulcer probably doesn't work hard enough'. When I looked at the remaining contenders, they all seemed like children.

The final seven all ended up having some sort of problem, so a consolation match began. This time the Russian medical team were doing the checks. I was asked to retake the test, and they didn't find any problems.

Akiyama was selected to undergo training as a cosmonaut candidate in Star City, the centre of the Russian space programme, alongside TBS camerawoman, Ryoko Kikuchi. Only one of them would be chosen to go to Mir.

TOYOHIRO AKIYAMA: When I went to Star City, everyone's roles were so clearly defined – the gatekeeper, the janitor, the secretary, the military men. It was just like the Aldous Huxley novel, Brave New World, where people are categorised as alpha, beta, gamma, epsilon, and so on, before being born. Apparently Star City is now on tourist routes, but at the time it wasn't even on the map, and the security was tight.

In Russian society, just being a cosmonaut candidate grants you high status, and I felt like I was a precious guest. Generally, the cosmonauts were ex-military, but they seemed much more educated than Japan's defence force. One of the cosmonauts, Anatoly Artsebarsky, was more like an intellectual than a military man. He seemed very interested in Japanese culture. I later learned that he graduated from the Air Force Academy and was a brilliant pilot. He was an elite among the elite. Making a good impression on the Russians was incredibly important. From the moment I arrived in Moscow, I was sure that I was being observed and evaluated. I wasn't naive enough to show my true self – I never showed an inch of my rebellious personality.

I had a typical late-forties salaryman lifestyle at the time. I'd be at work until late in the evening and then I'd go out drinking and socialising. As for smoking, it was a time when you could smoke at work. I smoked three to four packs per day, so thirty to forty cigarettes. Vladimir Shatalov, the head of the cosmonaut training centre, told me a few times to quit smoking. But when I first arrived in Star City in October 1989, I was invited to a party for cosmonauts at someone's house. I went there and as they opened the door, it was full of smoke. I thought I'd come to the wrong place, but I was welcomed in. I said, 'I thought we weren't supposed to smoke?' They said, 'Smoke until smoke comes out of your ears, drink vodka all night but feel refreshed in the morning – it's what qualifies you to be a cosmonaut.'

Right: Preparation for launch at Baikonur Cosmodrome.

After training for over a year, Akiyama was chosen for the Mir mission and was moved to Baikonur in Kazakhstan, the site of the Soviet space launches.

TOYOHIRO AKIYAMA: I thought Baikonur was close to Moscow, but we flew for hours. It was a desolate place. That's why the Soviets built their launch site there. When we arrived at the hotel, it seemed like it had been abandoned for months. But despite that, it's hallowed ground: Gagarin is like a god in Russia and Baikonur was where he first flew into space. That's why on the day of the launch cosmonauts do the exact same things Gagarin did before boarding the rocket. In the morning, you sign the door of the hotel room you stayed in. And because Gagarin took a piss before his launch, you unstrap your spacesuit and take a piss too [on the wheel of a bus on the way to the launch pad].

I woke up on the morning of the launch and went to the cafeteria, where I was served with the same breakfast as the day before. In Japan, you might get served with sea bream or something, like a celebration, but it was just as it always was – the same two slices of brown bread. We boarded about two hours before the launch and did all kinds of preparations and checks.

When we finally reached the countdown, for the first time I thought to myself, 'I've done it.' Until then, there are so many things that can cause a flight to be aborted.

Akiyama flew to Mir aboard Soyuz TM-11 on 2 December 1990.

TOYOHIRO AKIYAMA: When I entered Mir, I was exhilarated. I was then presented with bread and salt. In Russia, they give salt and bread to travellers, a tradition from the Silk Road. To do something like that on the space station is very creative. You're also supposed to receive a kiss from the most beautiful woman in the village but there were no beautiful women on Mir so I asked for a kiss on the cheek from flight engineer Gennady Strekalov instead.

I still didn't know how working alongside the Russians was going to go. That was the biggest concern that I had after arriving there. But I forgot all that when Strekalov and the rest of the crew gave me such a warm welcome. That was the moment where all my worries about cooperation with the Soviets went out the window and there was no unease. We'd arrived and the next stage was about to begin. I was delighted.

Although we used mock-ups of Mir in our training, my impression about the real thing was that it was really small, physically speaking. It was like a small factory with no gravity, where you had to work 24/7. It wasn't a place I would want to go back to again and again. The air is circulated and purified, but smell tends to remain. And the odour was terrible, like the locker room of a sports team. Specifically, like socks and sweat. But after a few days, I got used to it.

Above: Akiyama on his return from space.

Right: Akiyama wearing his Russian Sokol KV-2 space suit.

Akiyama endured a nasty dose of space sickness for the first few days of his flight, and described the uglier details to TBS viewers.

TOYOHIRO AKIYAMA: Because I was there as a reporter, I wanted to talk honestly and uninhibitedly about the changes my body was experiencing, that was part of my mission. I was not a member of that cosmonaut community. Rather than aspiring to be a hero, I knew that my job was to convey what was happening as truthfully as possible.

Everyone has their own mission. In my case, I was called a 'civilian cosmonaut'. Being a civilian and journalist meant that my mission was not to lie. I went to the space station to satisfy people's curiosity. A TV corporation had sent someone to space, so nationality wasn't a factor. If someone had asked me to sing the Japanese national anthem I would have refused, because in my mind I wasn't carrying the Japanese flag. In the case of Russia, most cosmonauts become national heroes. America is the same. I was disconnected from that context, and I think people understood that. The New York Times described me as a 'space anti-hero'. That's exactly what I was aiming for.

I think TV should have purpose. I hoped that by showing extraordinary footage of Earth (viewed from space) every day, I could imprint an understanding of the fragility of Earth's environment onto the subconscious of the Japanese people. I wanted to give people the feeling that they live on a precious and irreplaceable planet, not by reason, but by heart. To think, 'I never knew it was so beautiful.' I think humans tend to be convinced better by heart than by reason.

After staying on Mir for almost eight days, Akiyama returned to Earth on 10 December 1990.

TOYOHIRO AKIYAMA: I did feel like I wanted to stay a bit longer – not for an extended period, but after completing all the hectic work I did, I felt like I wanted time for myself, time to simply look out of the window and enjoy the view. The journey from Mir to Earth was short, only a few hours. The parachute deployed at 5,000 metres above the ground. There was a light jolt, and the G forces disappeared. When I returned to Earth, I just wanted to go and eat something delicious, have a cigarette and a beer. It felt like I'd returned to Earth as a ball of desires.

Akiyama is now in his early eighties. He lives a quiet life tending to a small farm in the Japanese countryside.

TOYOHIRO AKIYAMA: When I look back at my mission, I can't help but wonder if the TBS project was part of a bigger international strategy to get money into the Soviet space programme. I know from my time in Washington, there was great concern that if the industry fell apart, Soviet space engineers might migrate to countries the US saw as an enemy. In the late 1980's, America could not support the Soviet space programme directly, but I think investment from useful pawns like Japan was happily welcomed.

How do I feel about space exploration today? As human beings we have an innate desire to fulfil our intellectual curiosity and I think it's a good thing to go to space for the contribution it can make to science and to understanding our place in the universe. But if people are given totally free reign to pursue that curiosity, inevitably things like greed and profit come into the equation.

I have absolutely no faith that the leaders at the centre of International politics will act well. I'm certain there will be competition for resources and a race to claim territorial rights over other planets. Fundamentally, I think it is very important that we are all equal. But it's clear we have a long way to go before everyone is equal on Earth. If humanity cannot achieve peace in the Middle East or in Ukraine what can we possibly achieve on Mars?

I differentiate between the words "clever" and "wise" in English. I think humankind are clever enough to make rockets. But we don't yet have the wisdom to use them.

Left: The Soyuz TM-11 crew.

Right: The launch of Soyuz TM-11 from the launchpad in Baikonur, on 2 December 1990.

Left: Toyohiro Akiyama writing his signature on his door before he leaves for the bus.

Above: The Soyuz TM-11 crew during training..

Overleaf: A view over the northwestern part of the African continent.

LIVING IN SPACE

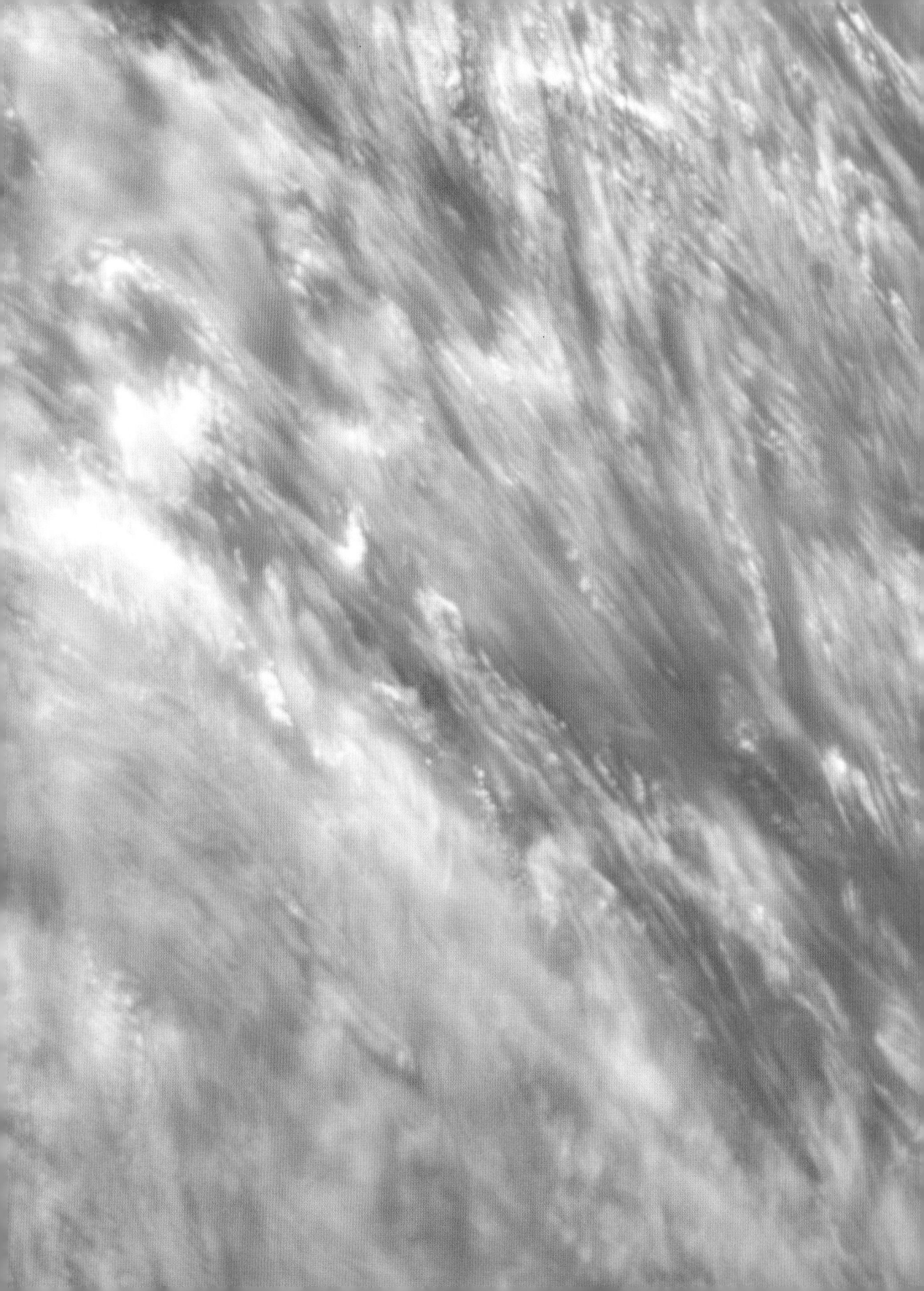

THE FALL OF THE USSR

As the USSR underwent huge political upheaval in the 1980s and early 1990s, eventually collapsing on 26 December 1991, the Soviet space programme continued. Aleksandr 'Sasha' Lazutkin was selected as a cosmonaut on 3 March 1992, and he later flew to Mir on Soyuz TM-25, spending 184 days aboard the space station. He is married to Ludmilla, and they have two daughters, Tasya and Zhenya.

LUDMILLA LAZUTKIN: I remember a few months before the Soviet Union fell, the tension was terrible. We were at home in Moscow, the children were at my parents' and Sasha and I drove to town. There were crowds of people on the streets and when we arrived at Lubyanka Square, we saw that they'd put some kind of cable around the neck of the monument to Dzerzhinsky [Soviet revolutionary and leader of the secret police] and had brought it down with a crane. We felt like this was vandalism. Not all the people were positive characters, but this is our history. And you need to know your history.

From that moment, everything started to disintegrate, in industry, in space exploration, and production began to dwindle. It was awful.

During this period, Russia started to undergo all of these transitions to capitalism – there was no food, terrible inflation. We were given vouchers for food, sugar, soap, vodka. Vodka was a very important commodity. We were given six cards, which was six bottles of vodka for the family, and this was a currency. Call any plumber, you have to pay with vodka. The inflation was insane, so the vodka was really great. We used it for absolutely all the services we needed. It was a very unpleasant period in the life of our country. There were crazy food queues; I absolutely can't understand how we got to a situation where there was no food.

Right: Queuing for food at a market in Sophia, Bulgaria, 1989.

122 LIVING IN SPACE

Below: Workers removing Felix Dzerzhinsky's statue in Lubyanka Square.

TASYA LAZUTKIN: In the Soviet period, cosmonauts were seen, if not as celestial beings, then stars, for sure. These people were considered simply incredible, which is what they are now too, of course, but back then they were revered, decorated. They were given apartments, cars. It was something that everyone admired. However, in the 1990s after the fall of the USSR, if anything, the profession generated irritation: 'Who do they think they are? What use do they have now? Who needs to fly to space when we have this situation in the country? What's it all for?' It was seen as over-indulgence.

Artsebarsky, the cosmonaut who hung the Soviet flag on the outside of Mir, was still on Mir during the 1991 coup.

ANATOLY ARTSEBARSKY: I was the last cosmonaut to be named hero of the Soviet Union.[3] Gagarin was the first, and I was the last. When Sergei Krikalev and I were on the station, we learned from radio enthusiasts in Australia and Europe that there was unrest in Moscow and a coup was happening. Mission control thought that we knew nothing about it, and they simply behaved as normal, not telling us anything. But we knew that Yeltsin had spoken to the people, and we even received his speech over the radio. We didn't know the details, just that there was unrest in Moscow. Now I understand that there were tanks and many people on the streets.

I copied Yeltsin's address to the people and sent it by electronic mail from Mir to all radio enthusiasts, because people didn't understand what was going on. It seemed to us that Yeltsin with the flag of new Russia was such a proper man, a proper Communist, only later we came to learn that none of that was true.

We had a Soyuz ship, and we could have returned at any moment if anything happened. But we never even gave that a thought. We just waited for the situation to resolve itself. Personally, I thought that it would all take longer, but everything appeared to have been sorted out in two days.

The USSR collapsed unintentionally, because people like Yeltsin, Shushkevich, Kravchuk from Ukraine, came together and decided to cancel the Soviet Union. They never asked the people. The people voted for the Soviet Union to be kept together. But the people in power voted to tear it asunder. Those actions were wrong. I am sure that it would never have happened without the Western idea [of capitalism].

I love Russia. I loved the USSR. It is not something you can describe in words. It is in my heart. What happened was not something I could accept. I remained a Soviet citizen for all time, and that is why I raised the Soviet flag.

When I was training to be a cosmonaut, I was thinking about raising our Soviet flag [one day], because the Americans had raised theirs on the Moon. They are patriots, and they hang their flags everywhere, even on their homes. As a patriot, I was proud that our country could build such a unique space station on its own, and that was why I put up the flag – to mark the achievement of the Soviet Union.

I placed the flag on the Sofora rig[4] [of the Mir space station] when it was still in the horizontal position. Then we raised it up. After that, I climbed all the way up the rig to secure the flag at the very top. When I was looking down the rig at the station from a height of 15 metres, I was terrified. The fact that the Earth was 400 kilometres further on was totally fine! Not an issue! I thought that I would fall, flag in hand, and fly round the Earth like that and never return. Since then, I have a fear of heights. It was complicated work, but we managed it.

The flag eventually disintegrated from serious cosmic radiation and drops in temperature. It fell apart just like the Soviet Union. Before that, all the Soviet flags on Earth were taken down, but they couldn't reach the one up there on Mir. It remained up there proudly for a whole year as a symbol of the mighty Soviet Union. That is hard to think about. But what can you do? That time has gone.

---

[3] The title 'Hero of the Soviet Union' was the highest distinction and honour awarded to Russians.

[4] One of the largest external component of the space station, the Sofora rig is a large 14m long scaffolding-like structure consisting of 20 segments, mounted of Mir's Kvant-1 module.

Left: Anatoly Artsebarsky's crew photo from Soyuz TM-12.

Above: Artsebarsky signing his crew photo.

Overleaf: Mir with the Soviet flag on the Sofora rig.

LIVING IN SPACE 125

THE SHUTTLE-MIR PROGRAMME

With the newly formed Russian Federation experiencing political turmoil as the country transitioned away from Communist rule, the USA saw space as an opportunity to build bridges between the two countries.

FRANKLIN CHANG-DÍAZ:
Challenger took all the bandwidth of our brains, and we were not really concerned with what the Russians were doing. We were focused on how to deal with this terrible tragedy and how to get back flying. The USSR, meanwhile, was working on the Mir space station, which only became a point of attention later when it became apparent that the Soviets were spying on us. Maybe we were using satellites instead of astronauts to spy, but everybody was spying on everybody else, and that's what was driving the USA to build Freedom, a US-only space station that was going to be the answer to the Soviet's space station. The Russians had Mir, and the Americans had the shuttle, and they could fly the shuttle any time, but they didn't have a space station to go to. However, space station Freedom was hanging by a thin thread in Congress, and there was really very little support for it as it was too expensive.

The collapse of the Soviet Union precipitated a lot of things, including finding a way for the whole scientific infrastructure of the USSR to not disperse and go to rogue nations and create a lot of havoc. There was a fear that Soviet scientists were going to basically spill over into other nations that were not interested in competing in space, but were interested in developing weapons.

This was an opportunity to bring together the two programmes, one with a space station already operating and the other with a space shuttle that has the capability to deliver lots of stuff to it. That's how the Shuttle-Mir programme was born.

The first flight of this programme was an exercise to judge the ability of astronauts and cosmonauts to work together – a flight to break the ice. It was my fourth flight, and I have no idea why they picked me to go. Maybe they needed a Hispanic to teach the Russians how to speak Spanish! I never really felt Russia to be the enemy, whereas many of my colleagues being ex-military saw them that way, plain and simple.

When we went to Russia for the first time, in the spring of 1993, we received a very cold reception. We got there and their attitude seemed to be, 'Why are you here? We don't want you,' and ours was, 'Well, we don't want to be here either, but they've told us we have to work together,' so we had to try and figure out how to do it.

Above: The STS-60 mission patch. STS-60 was the first Shuttle flight to approach the Mir space station in orbit, while the later STS-63 mission made a rendezvous with Mir.

Right: The STS-60 crew: (clockwise from bottom left) Kenneth Reightler, Franklin Chang-Díaz, Ronald Sega, Sergei Krikalev, Jan Davis and Charles Bolden. Sergei Krikalev was the first cosmonaut to ride NASA's space shuttle.

LIVING IN SPACE 129

Left: An in-flight portrait of the STS-60 crew, floating in front of the US and Russian flags.

Above: The 18th launch of Space Shuttle Discovery on its STS-60 mission to Mir.

Michael Foale was selected in NASA Astronaut Group 12 on 5 June 1987, and he came to play a significant role in the Shuttle-Mir programme.

MICHAEL FOALE: Russia was really, really good at putting people into space and letting them live there for months at a time. Yes, the US had done that, but only a couple of times in the Skylab programme in the mid-70s. Really, Russia had developed all of the expertise and had all the knowledge about what the effects of living and working in space were on humans. At that point in the mid-90s, Russia was clearly the world leader for living in space.

So, a programme was invented to start Americans and Russians on the path of building the International Space Station – the Americans called it Shuttle-Mir and the Russians Mir-Shuttle. That programme was Phase One of the plan. Phase Two was building the ISS – but Phase Two wouldn't happen if the Shuttle-Mir programme was unsuccessful. We needed to prove we could work together.

Although representing NASA, Michael was a British citizen. Having dreamed of going to space from a young age, he decided that science was his best route into the corps.

MICHAEL FOALE: I applied to Cambridge to study physics and was accepted. Around this time, in 1978, the US space shuttle programme was being talked about in the media: they were going to fly non-pilot astronauts, and they were going to be scientists. This was brand new. I thought, 'Whoa, that could be me.' I set all my energy on going to the USA and moving to Texas, because that's where astronaut selection and training happens, and because 'Houston' was the first word Neil Armstrong said when he landed on the Moon.

Once in Houston, I was hired by McDonnell Douglas, which was based just outside of the Johnson Space Center, to work on space shuttle navigation. I didn't know much about space navigation, but I was a physicist and thought I could figure it out. I was pretty arrogant about that sort of stuff. I enjoyed it and had good colleagues. And sometimes I'd go to the space centre in the middle of the night, sit in mission control in a back room and monitor the navigation data coming from the shuttle. It was exciting being right there where the action was.

Left: Michael Foale talks to amateur radio operators on Earth via the Shuttle Amateur Radio Experiment (SAREX), during the STS-56 mission, April 1993.

After joining NASA and being a crew member on two space shuttle missions, Foale was selected for flight STS-63, the second Shuttle-Mir mission, which launched on 3 February 1995.

MICHAEL FOALE: I was getting ready to do my first spacewalk when we were told that we were going to have a Russian on our flight, Vladimir Titov, who would not speak English well. He was assigned late in our training flow, but he was an experienced cosmonaut who had been on the Mir space station. Following an idea that President Bill Clinton had developed, the people above me had gone to Moscow and made an agreement to have two Russian cosmonauts come to the United States to fly on the shuttle: Sergei Krikalev, who is currently head of Putin's space programme, and Vladimir Titov, who had been blown up on a rocket and survived.

The international aspect of the space station was not in our minds at the time. Today, I think all of us would agree that the international aspects of the International Space Station are fantastic. It's one of the few areas where Russia and the United States and the West are still collaborating. But, at that time, we saw it as a negative. Nonetheless, we were good soldiers. We wanted to fly in space again, and I wanted to do my spacewalk, so none of us were going to object. We were respectful to Vladimir Titov, who practised his English with us.

I flew as a flight engineer that time, and I was responsible for helping the shuttle get close to Mir, because that was our new goal. Someone asked, 'Can we dock with it?' We didn't even know what this thing looked like. We were told, 'No, you're not going to have a docking adapter. You're just going to fly up to it for about ten minutes and wave.'

Because we used Imperial units, and the Russians used the metric system, we didn't even know if our communication systems would work to allow us to bring the space shuttle up close to Mir. But if we were going to build an International Space Station, we needed to know how to do this. And it was very cool. Vladimir turned out to be a great cosmonaut to fly with.

Below: Rendezvous and approach of the orbiter Discovery to the Mir space station, 1998.

Left: Cosmonaut Polyakov watches Discovery's rendezvouz with Mir.

Below: Mission specialist Michael Foale (with the red stripe on his leg) is pictured on the SRMS arm, preparing to grab Spartan 204 as payload commander Bernard Harris Jr. looks on. Part of STS-63 mission's spacewalk.

LIVING IN SPACE 135

Jerry Linenger, who was selected in NASA Astronaut Group 14 on 31 March 1992, was also chosen to join the Shuttle-Mir programme, and he moved to Russia to train in January 1995.

JERRY LINENGER: I went to the Naval Academy, because it had more astronauts than any other school. [I then went] via a special programme directly to medical school and became a Navy flight surgeon. I did a lot of flying in jets stationed overseas in the Philippines, and for a couple of years, I worked for a three-star admiral, which was an honour. Then, eventually, I got into the astronaut programme, after I had been fortunate to serve our country for twenty-two years in the US Navy.

I was running a sports medicine group at that time, and I was in my office doing research when I got a phone call from Johnson Space Center. Mr Don Puddy[5] said, 'Jerry, if you want to join us, you've been selected to the astronaut corps. Come on down to Houston in August.' I threw the phone down and started screaming at the top of my lungs! When I picked up the phone again, he said, 'We haven't made the official announcement yet. The press conference is not till tomorrow, so please tell no one.' I hung up the phone, and, of course, everyone in the research centre was wondering what was going on and what all the yelling was about. I picked up a piece of paper and said, 'Man, the test results came out really good. I'm just so excited!' They all looked at me like I was losing it, but I managed to keep my mouth shut. Loose lips sink ships, as they say.

While I was an astronaut candidate in training, they brought up at one of the weekly meetings that we were going to be doing some cooperative flights with the Russians and to have a think about whether we'd be interested. They had a lot of experience, and the International Space Station was on the horizon, so we needed to start getting some international cooperation going. I was a new astronaut, and if there were flights to space coming up pretty soon, that was something I was interested in. My medical background also meant I was interested in human physiology, and I thought it would be fascinating to see how my body reacted to living in space for months at a time. So, I put my name into the mix for the 'Russian thing', as we called it.

I was called into Hoot Gibson's office,[6] and he said, 'Jerry, we'd like you to do the Shuttle-Mir thing that we talked about. You better get ready to run, because you have to learn the language and everything else.' It was not a 'yahoo!' moment. It was more of a thoughtful moment, thinking, *There was going to be a lot of change.*

There was a whole lot of logistical stuff to consider. It would also mean learning Russian and working with Cold War enemies. It was a tough undertaking preparing for the mission to Mir in a very compressed timeline. I had to work very hard, as I did not want to be the weak link in the chain.

Star City was a community of people that had worked in the space programme pretty much forever, so some of the trainers were older men who had been doing this for the last thirty years. I understood that when they were trying to teach me something, they didn't want to be my friend. There was a little bit of, 'Why is a US naval officer sitting here in front of me, and what's he doing in Star City? He's one of the enemy.' They had a job to do, so it wasn't horrible, but there was tension, no doubt about it. I decided I needed to put that away and integrate and become a cosmonaut.

The Russians were very kind and open to my wife, much more so than they were

---

[5]   The head of the Flight Crew Operations Directorate at the Johnson Space Center following the departure of George Abbey.

[6]   Robert 'Hoot' Gibson was chief of the Astronaut Office from 1992 to 1994.

toward an American in the middle of their space programme. They had a lot of pride, and they did not like the idea that the only reason their programme was still going was because of US help. The American astronauts represented that help, and I would say that we were probably not the most popular people on that Star City base.

The family members of the cosmonauts had to start commuting into Moscow to get second jobs to help bring in income. And when I was in space, one night after I went to bed, floating on my wall, I had to get up again to do something I'd forgotten to do, and I saw my two cosmonaut crewmates doing a commercial for milk or something. They were tired, but they had to do it to bring money into the Russian space programme. They were almost embarrassed and wanted to do it after I'd gone to bed.

Top: Jerry Linenger in astronaut training, 1993.

Bottom: Star City, Russian's main cosmonaut training centre on the outskirts of Moscow.

Michael Foale and his wife Rhonda moved with their two children to Star City on Thanksgiving Day in 1996.

RHONDA FOALE: Mike called me from Russia on a separate trip and said, 'Oh my gosh, the secretary told me that we have to move to Russia next month.' I probably went numb thinking about all the things I had to do in six weeks. I'd been mentally preparing that we were likely to have to move to Russia for a few years, so I was excited to try a completely new place and learn about a different culture and history, but I was apprehensive as well. Yeltsin was president, and I knew about the oligarchs and criminal turmoil. But I thought, *'Let's do it and see how it goes.'* I wanted to be there for Mike.

[We arrived in Russia in] late November and it was very gloomy, a bit slushy, and the airport was very depressing. We had to drive for an hour out into the country to get to this little hidden village where the cosmonaut training centre was. It was meant to be a secret city, so it was very grim and crumbly. They showed us the little room that we'd be living in for a while. There was barely enough space for the four of us in there. It felt like going back to the 1950s. It was very different to sunny Houston on Galveston Bay, but I knew I had to get through this and figure it out.

About six weeks later, they finally got these homes built that were supposedly to American standards for the astronauts to live in. It didn't even seem like Russia to me. I was picturing St Basil's Cathedral and beautiful countryside and *Dr Zhivago* and quaint little wooden houses, but where we lived wasn't anything like that. It was just a planned town built in 1962 with a double concrete wall around it. We had to have a little passport to get in or out, and we weren't allowed to drive cars. We were just stuck in Star City.

When we moved there, Russia was still in turmoil from the dissolution of the Soviet Union, and all you could really depend on buying at the store was alcohol, cigarettes and hard candy. I didn't recognise any of the meat. One time I asked for some carrots, and they handed me this big plastic bag full of black dirt and I said, 'Are there carrots in there?' There were, so I took it home. I was really jealous of the Germans,[7] because it was just a short flight for them to go home, and they'd come back with suitcases of food.

---

[7] In an extension of their Interkosmos programme, the Russian space agency signed an agreement with West Germany to send their first astronaut to Mir in 1992.

MICHAEL FOALE: Jerry was about as uptight as I was being in Russia and trying to learn Russian and eating in a separate dining hall so the microphones could pick up what our conversation was at lunchtime. All the foreigners were put together, and the cosmonauts who we were going to fly with one day were all in their uniforms around their own tables. So, I got to meet these European astronauts because they were also spies as far as the Russians were concerned.

Rhonda in particular liked to have the parties, and we'd play good, loud Russian disco music and other European music. We'd invite all the European astronauts and their families, and we'd all dance in the kitchen.

My first memory of [cosmonaut Sasha Lazutkin and his wife Ludmilla] was going to a circus – I think they invited Rhonda, me and the children to go with their two young daughters. They were just incredibly calm, polite and hospitable. We then followed up by inviting them to our house. They also came to one of our parties, where we would invite Russian cosmonauts who were around and willing to come, and we'd be dancing in the kitchen again. It became such a fun thing to do that later on, we converted the basement of one of the cottages into a bar, and, much later on, I supervised the importation of a pool table and gym equipment to make a bar that everybody in Star City wanted to go to. We even had Tom Hanks at that bar.

RHONDA FOALE: Halloween is not a Russian holiday, but it's a big, fun thing to celebrate in America. One of the American astronauts who was pretty new in Star City and wasn't yet familiar with Russian customs, said, 'Hey, let's have a Halloween party.' He told the Russians to wear costumes – however, in Russian 'costume' means 'business suit'. Trying to rustle up Halloween costumes in Russia is pretty hard, so several people went as Roman senators, you know, with bed sheets, but some people made the most of the fact we had medical supplies – bandages and the like. So we arrived at Dave Wolf's apartment door with some Russians wearing their business suits, me in a sheet. We knock on the door and Dave opens it – his costume was a prosthetic thing on his head that looked like a golf ball had smashed into his skull with blood everywhere. As soon as he opens the door, all the Russians shriek and call emergency services, and we start laughing and laughing. It was a bit of a shock for them, but it was a really fun party!

Linenger flew to Mir on STS-81 on 12 January 1997. The Russian cosmonauts Aleksandr Kaleri and Valery Korzun had been there since 18 August 1996.

Below: Jerry Linenger shaking hands with commander Valery Korzun on arrival on Mir, 1997.

Right: A view from Mir, looking down to one of its solar arrays at Space Shuttle Atlantis (STS-81) docked to the station through a module in its open cargo bay. Another of Mir's solar arrays is visible in the top of the image.

JERRY LINENGER: We did some work on the shuttle before we got to Mir, but our main goal was rendezvous and docking, and the main cargo was me. They were then taking the American John Blaha [back to Earth].

Docking in space is not easy. You're going 17,500 miles an hour, and Mir's doing the same thing. You've got to catch it, and it took about a day and a half to get to where we wanted. The target is a big circular ring on Mir, and we've got an identical one on the shuttle. The key is to line up those docking rings. If you go up too fast, you have a collision. If you go up too slowly, you just sort of bounce off.

Mike Baker, the shuttle commander, brought us in, and the people on Earth told us the alignment looked good. The mechanical latches closed, and we equalised the pressure. After a few tests, we opened the hatch, and I flew on board and said, 'zdravstvyite' (hello) to my new Russian crewmates [Aleksandr Kaleri and Valery Korzun].

Overleaf: Night launch of Space Shuttle Atlantis (STS-81) to Mir on 12 January 1997.

LIVING IN SPACE 141

Meanwhile, Aleksandr (Sasha) Lazutkin was preparing to travel to Mir and join the crew there.

LUDMILLA LAZUTKIN: We did the traditional visit to the Mausoleum in Red Square, and we laid flowers for Korolev [Soviet aerospace engineer] and Gagarin, and took the official photos. I went through the whole gamut of emotions, because I realised that there was nothing else to wait for – Sasha was flying into space. We were walking in Red Square, and I wanted to shout about it and let everyone know that here they are, the future cosmonauts. Well, of course, I wasn't allowed to do that.

At the end of January, the children and I went to see them off in Star City [before they departed for Baikonur]. I remember that [the farewell ceremony] was about to begin, and I started to feel physically unwell. The stress, fear, panic, I couldn't make it go away. I had to get myself together because they were making toasts to see the crew off. In the official photos of the farewell, I have my glasses on because I had tears in my eyes. Just fear, fear, fear.

Then we were sending them off, and the children were saying goodbye to their daddy. It was very touching, and we were anxious and excited. Sasha said, 'Don't worry, we still have two weeks before the flight. We're going to Baikonur first. We're not leaving tomorrow.' But I knew I wouldn't see him again [before the launch]. He'd already flown away in my mind. I would only watch him on TV later.

ZHENYA LAZUTKIN: My understanding of time was rather elastic. Yes, I knew that Dad was in training and that he was a cosmonaut, but it didn't somehow fit into the big picture. So, when the time came for him to [go to Mir], it was a huge shock for me. I'm in the photographs standing next to him, bawling my eyes out. That reality seemed to have been dumped on me. And that six months really did seem like an eternity.

TASYA LAZUTKIN: I can't say that we were joyous. We were happy that Dad was flying into space, but instead of sitting in a close family group, spending quality time together before the mission, we were forced to be among strangers who said things that no one was really listening to. There were protracted speeches, and you needed to hold yourself together. So, the point when they got on the bus triggered those powerful emotions. You'd been holding those emotions inside, and you didn't need to any more. I wanted to grab him and hold on.

LUDMILLA LAZUTKIN: As they were leaving, Larisa, the wife of cosmonaut Vasily Tsibliyev, and Tamara Globa[8] were standing next to me, and Tamara was crying for some reason. I knew why I was shaking, and I knew why Larisa was upset, but I thought, *'Tamara and Sasha are friends, but to get really upset like this ...'* She pulled herself together and said, 'They're going to have a very difficult flight. God willing, they come back.'

---

[8] Tamara Globa is a famous Russian medium/astrologer.

Left: Sasha Lazutkin training in the neutral buoyancy lab at Star City, whilst his daughters sit at the training console.

Right: Lazutkin family portrait.

Below: Soyuz TM-25 launch.

Right: Lazutkin on board Mir.

SASHA LAZUTKIN: When I started my training for the launch [after saying goodbye to my family], my commander, Vasily Tsibliyev, asked me if I was baptised or not. I told him I wasn't. A day before the launch, the woman who served our breakfast came up to me and asked, 'Sasha, are you baptised?' I again replied that I wasn't. She said, 'You should go and get baptised.' I said, 'I have to come to that decision myself. I'll do it after the flight.' And she said, 'You need to do it before the flight.' I said, 'I'm going into space tomorrow. I can't go anywhere now, and there's no church nearby.' She said, 'It's all agreed. The priest is on his way.'

That was how my baptism took place. Why it was needed at that moment before the launch, I don't know. When we returned from the flight, someone said that I had stayed alive because I'd been baptised. Perhaps there is some truth in that, I do not reject the notion.

On the launch day I thought I should be feeling worried, but I was actually quite calm. We gathered all our things and travelled out to the launch pad. They put the spacesuit on me, and we walked to the rocket. I realised that these were my final steps, but I wasn't worried. For me, that was a revelation, to be honest,

because I didn't expect that from myself. I was checking my pulse. It was in the region of sixty-six to seventy-two – quite normal. I asked myself why that was. Even when the engines fired up, I took that completely calmly. And

then the rocket took off and the forces kicked in. Not a single worry or concern. The first abnormality was when we were flying, and I understood we were at a considerable altitude. The second stage of the rocket became operational and suddenly the G-force was gone, and I felt as if we were tumbling. I thought,

'*Why are we doing this?*' Then I heard the voice from Earth, and it said, 'The second-stage engines have been shut down. The third-stage engines have started.' I felt the G-force increasing.

Next, we were congratulated for entering orbit, and then solar batteries[9] opened out and the antennae assumed their positions. There was a brief setback – the voice from mission control informed us that one antenna hadn't opened fully. This was our first 'non-standard situation'. When I was still preparing for the flight, I'd asked the cosmonauts how many non-standard situations to expect in that six-month period. Space equipment is highly reliable, and they said to me, 'About three non-standard situations. Perhaps four. Five at most.' I was flying for the first time so I was afraid of those non-standard situations. Normal operation was not something I feared, but non-standard situations – I didn't know how I'd react. Would I be able to handle them? And now the first non-standard situation had occurred in the first minute of the flight. I counted off the first finger on my hand. Everything was fine.

That night [on the Soyuz], an incident warning signal came on. Some other

---

[9] Like Mir, the Soyuz spacecraft also used solar arrays to charge battteris to provide electrical power.

device had broken. There was nothing terrible there, but that was the second non-standard situation. Then, on the second day, we were flying towards the station, observing the process, because the craft was being controlled by computer. We approached very close, with just two metres to go, when suddenly the vessel came to a stop. An incident warning light came on, and the vessel began withdrawing in accordance with its own program. The commander and I switched to manual, took over the controls and docked with the station. And I thought, *'Well, that's three!'* The mission had only just begun, and I thought it would all go well, because three non-standard situations had occurred, and we had six months still to go.

When I arrived at the station, I was a little afraid of it. It was so big! And I thought, *'This station is now my responsibility.'* My life in weightlessness had also begun. The body growing accustomed to it is basically a painful process. You have a headache, because the blood is always going to your head. You are forever feeling nauseous. Your spine hurts, because in weightlessness the body is stretched.

Everyone overcomes that pain in their own way. I was one of those who suffered. The next day, it went from bad to worse, and I got to the point when I realised I couldn't go on. That was the state I was in. I'd heard people say, 'No one has died from it yet,' but I thought I was going to be the first.

That evening, we were all round the table to have our dinner. I hadn't told anyone I was feeling rough. I said to the others, 'Guys, I'm going to bed.' And they replied, 'Sure, off you go.' I was convinced I would not make it to the morning. I got into my sleeping bag, and before closing my eyes, I cast my gaze around the module because I was saying goodbye to it. I had reached that point. I remember that I was calm; I'd made my peace with it.

The next morning, I woke up and my first thought was, *'I am alive?'* With a question mark. I slowly floated out and felt that all the symptoms had simply gone. At that moment, I realised I would live. And then I had this emotional explosion: 'I am alive!' I said to the guys, 'I was born today.'

LIVING IN SPACE 147

THE FIRE

Below: Scorch marks on the outer shell of Mir.

SASHA LAZUTKIN: The day of the fire was 23 February – a public holiday for us in Russia, Defenders of the Fatherland Day. Early in the morning, the Patriarch came over the comms – the Patriarch of all Russia, Alexy II – and congratulated us on this day. That was how the day began. At that time there were two crews on the station, meaning six people [and more oxygen needed]. One crew was nearing the end of its time and we were only starting ours. We were having dinner when Commander Valery Korzun told me that we needed to deploy an oxygen charge.

This is a perfectly normal operation: a chemical reaction begins, and the oxygen starts being released. I went over, deployed the canister in the installation, and the reaction started. But somehow, the oxygen charge caught fire.

Everything burns where there is oxygen. But it wasn't any substance that was burning, it was metal. The burning temperature of steel is 1,200 degrees. The oxygen charge canister stood next to the outer shell of the station, which was made of aluminium. It would only need some 800 degrees and it would turn into fluid – that would have been a terrifying outcome of the fire. Putting it out was problematic because the extinguishers that released water were no good since water turns to vapour at 100 degrees. And it was 1,200 there.

The alarm went off and the command was transmitted all over the station.

JERRY LINENGER: The last thing I remember is Sasha going into a different module and inserting the canister, then I flew away to another module to enter some data on a laptop computer. Then I heard the master alarm going off. I pushed off to turn the corner and return to the base block, but before I could get there, Vasily came flying around the corner fast, and we almost had a collision. I said, 'Is it serious?' And he replied, 'Yes, very – fire.' I could now look down the length of the base block, about the size of a small school bus, and see that it was rapidly filling with smoke. I could also see where the flame was coming from. It was a big flame, about two or three feet in length, blowtorch-like in intensity, with sparks flying off the end. I'd done firefighting before, but I had never seen smoke spread like it did on that space station within the first thirty seconds or so. I knew I needed to get a respirator on in order to survive.

I went back to where I knew an oxygen canister was located and yanked it off the wall. I was starting to really need oxygen, and it was getting to the point where I couldn't see the five fingers in front of my face. I put the rubber mask over my head and flicked the lever to activate it, breathed in – but I got nothing. Plan B was to go find the next respirator, so I started feeling my way along the bulkhead, trying to get to a respirator that I wasn't even sure was there.

A lot of thoughts were racing through my head. The first one was: find your way along the bulkhead, get a respirator, get the thing working, start breathing and go fight the fire. But the second one was a more human, emotional one. Out loud, I actually said goodbye to my son John: 'Sorry, I'm letting you down. Looks like I won't be making it back. I'll be watching over you.' I then looked around and said, 'Wow, what a strange place to die.' I was 250 miles above the Earth with some Russians, things floating everywhere, smoke everywhere, and I sort of accepted that was it and thought, *It's part of life. I just wish it wasn't happening so soon.*

I finally found a second respirator, yanked it off the wall, put the mask over my head. I got oxygen and hyperventilated for the next thirty seconds or so to get my blood oxygen level back up. I then screamed out, 'We're going to get that fire out. I'm going to see my son again.'

SASHA LAZUTKIN: For me, everything went so slowly, I was in shock. I kept thinking, *This is made from non-combustible materials. It does not burn, and yet it is burning. Why?* I was fighting the fire, and the alarm sounded loudly throughout the station. Valery flew to me, took the extinguisher from me, without even asking, just took it, and shouted to everyone, 'The fire extinguishers!' And everyone, the remaining five men, scattered to all the modules in the search for fire extinguishers, and they started bringing them over. We worked on putting out the fire; the station was full of smoke. Again, Valery Korzun shouted once more, 'Gas masks on!' And everyone scattered to put on their gas masks. We put that fire out pretty quickly, in about fifteen minutes, I reckon.

JERRY LINENGER: I headed back toward the module. Valery Korzun, the experienced commander who was already on board, and I decided we were going to fight the fire, and we sent the new crew, Sasha and Vasily, to get a Soyuz capsule ready for evacuation if need be. There were two Soyuz capsules, but, unfortunately, one was located on the other side of the flame. In order to evacuate, you'd have to go through the fire, and that did not look survivable. So, we had three of the crew getting ready to evacuate in one of the vehicles, and the other three, the more experienced guys, fighting the fire. You needed the professionals doing the things that you have to do in a crisis situation.

Because of the fire's location, we could only get one fire extinguisher at the fire at a time. So, I activated an extinguisher and passed it to Valery. He fired it and was thrown back, like a thruster – physics 101. I worked my way into the connecting tunnel and sort of wedged my legs in there and wrapped my arms around Valery to stabilise him so he could operate the extinguisher. But we really had no chance against that flame: we had a water-based fire extinguisher so it was generating a lot of steam, which along with the smoke really added to the blackout conditions. We quickly changed strategy, and tried to direct the fire extinguisher at the bulkhead to keep any secondary fires from starting. We very quickly realised that the hull was made of quite thin aluminium. I was also seeing splatter on the far bulkhead – it kind of looked like balls of wax. As I squinted and looked harder, I realised it wasn't balls of wax; it was molten metal. If [it breached the] hull, we were going to have rapid decompression, quick suffocation and not a lot of options.

I was heading for the fourth fire extinguisher, fourteen minutes into the fire, when Valery screamed out, 'Jerry, fire's out!' There was a big sigh of relief, but only for a second, because the doctor in me took over, and I realised we were still in big trouble. The flame had essentially used up all of the oxygen [in the canister], and the only thing keeping us each alive were the little respirators, as we still had an unbreathable atmosphere. So, I told my crewmates to go to the end of a distant module and slow their metabolic rates, making every breath count. I joined them and pulled out a medical kit, ready to intubate somebody if they went into respiratory failure from the smoke inhalation. Then I just closed my eyes and sucked the oxygen as slowly as I could.

Finally, Valery's oxygen ran out, so he took off his mask and sniffed the air. He was able to breathe. What really helped was the steam floating in the air and hitting the cold hull and condensing, as it took a lot of the smoke and toxic products along with it. Eventually we took rags and our old, sweaty clothes to mop up the walls, and they were just black with debris. The Russians would later say that it was a great life support system that filtered air wonderfully, but I'm not so sure. We spent the next twenty-four hours or so stabilising things, trying to get the oxygen levels back up, wiping things down. We actually all stripped. It was kind of hilarious – six naked guys out there [in space]. After about forty-eight hours, I went to my wall, strapped myself to it, closed my eyes and slept like a baby.

When I woke up, I thought, *'Thank God I'm alive.'* I was still very concerned about what we were breathing in, but I had to communicate to the Earth in coded language, if you will, because they [my Russian crewmates] didn't want to talk about the fire. I said, 'What do you want me to do? I'm the medical person up here. What are we breathing in? What noxious chemicals are in the atmosphere now? And how should we counter that for our future health?' So, I had a lot of concerns for the following week after the fire.

SASHA LAZUTKIN: Later, once the fire was out, we discussed and assessed everything that had happened. It was then that we all felt genuinely terrified, because we understood that if it had continued to burn like that, the station hull would have burned through. And quite easily, too, because the hull was pretty thin – about three millimetres thick. A hole would have formed, and we would have died in a matter of seconds, simply because the air would have been released into space. If anything, that struggle for life brought us closer together. We became a tighter-knit group.

Back on Earth, they tried to work out why the fire had happened. It was completely improbable: an oxygen charge was a wholly reliable thing and down on Earth they couldn't set it alight just like that. As we were told, one charge in a million could have caught fire. Now, however, they were sure that they could try to set light to another million charges and there'd be no fire. But there had been a fire. I was the person who deployed that charge and it went up in flames pretty much before my eyes. I couldn't understand why it was burning when everyone had told me that it could not burn. I was in a state of shock, really.

JERRY LINENGER: Vasily and Sasha were both absolutely blameless. All Sasha did was activate an oxygen canister that we needed to breathe. But it's easier to blame a person and say they are the problem, not our system. Mir was the only show in town for the Russians, and they had to do their best to keep it going.

Above: Jerry Linenger using a respirator after the fire.

LIVING IN SPACE 151

LUDMILLA LAZUTKIN: I didn't learn about the fire from some special people calling me and trying to calm me down. No, I heard it on the radio. I was at work, and I just heard on the news quietly in the background that there was a fire on board Mir, and there was no communication with the crew. Then they moved on to other news. We were overwhelmed with news in 1997. I've had lots of shocking moments, and this was one of them. I left the office without saying anything and started to think what to do? That's when I realised I couldn't make a phone call to Star City, because I didn't have anyone's number, and I didn't know who to call. It was a nightmare. I went to my boss and said, 'I'm leaving. I'm in no state to do any more work today.'

I went home and decided to call Larisa, Vasily Tsibliyev's wife. She was great. She said, 'I don't know anything, but they say everything is fine. They were able to get in touch on the next orbit.' The journalists weren't given this information, but on the inside, the specialists, they knew what the situation was, and they'd called Larisa and told her something. She said, 'Let's calm down and wait. They're alive. For now, they're alive.' The orbit takes an hour and a half, and you're sitting and waiting – will something happen or not?

I honestly don't remember how long this nightmare lasted. It felt like for ever to me. They broadcast the same thing on all of the news channels, and sometime in the evening they showed the footage of the crew wearing gas masks. They were there, moving at least.

I hid my fear. What else could I do with it? In some moments, if no one was there, I'd cry. When you are trying to hide things from the children, at some point you contract like a spring. Then you find a way: you go to your girlfriend's place, you talk to her for half a night, you cry it all out, you have a glass of wine, you relax, you come home, you smile again, and you continue to live and work.

TASYA LAZUTKIN: No one spoke specifically with us about it, but we overheard things. When I was little, Dad had told me what I should do if there is a fire. We lived on the ground floor, and he said, 'You open the window and jump out if need be. You throw your sister out – she will come to no harm. We're on the ground, so you jump.' But out there, where would you jump?

Above: The remains of an oxygen generator which had caught fire on Mir.

Right: Mir, backdropped against Earth, taken from Space Shuttle Atlantis following undocking from the station at the end of STS-71, 4 July 1995.

Foale flew to Mir on flight STS-84 on 15 May 1997, to take over from Linenger.

MICHAEL FOALE: I'd been in Star City training for a few months when Jerry had sent a report describing what turned out to be an incredibly serious event, in which a fire had almost burned through the hull. His final words were something like, 'NASA should consider not continuing this programme.' I remember looking out the window thinking, *'Is this going to stop me flying?'* I thought Jerry had written as objective a report as he could. He put an opinion in there, but he didn't try to double guess what management would decide.

My take on risks and doing something that people perceive as dangerous is that lightning doesn't strike twice in the same place. I felt sure that there wasn't any Russian who was going to allow any more faulty canisters to be used in space.

I didn't have to say that I [still] wanted to go. I had put so much time and effort into learning Russian, and I was committed to going, even though I knew it was going to be the most difficult space mission of my career, because of the language barriers and because my role on Mir would be so diminished compared to my roles on the shuttle. There was no spacewalk planned for me. There were only experiments to do, and, based on what Jerry had written, there was some tedium involved. Nonetheless, even after his report of the fire, I thought, *I want to do it.*

JERRY LINENGER: Mike Foale coming aboard was a glorious moment. I was very glad to see him, and I tried my best to brief him and make sure that he was going to be safe, that he understood the reality of things on Mir and not the story that he might have gotten down on Earth. It was also good to have my relief here, as I was ready to leave Mir.

Saying goodbye to Vasily and Sasha was definitely harder than saying goodbye to Mike, because they were my crewmates for four months. We went through some tough times, and they were my Band of Brothers.

Space travel is an incredible experience, and I am so thankful and privileged that I was able to do it. Does it have some tough consequences? Does it take mental fortitude? Does it play with your psyche when you're isolated like that? Absolutely. But on the other hand, there's something great about being tested and coming out the other end whole and saying, 'Man, it's amazing what I can handle, and, more broadly, what human beings can handle.' We shouldn't sell ourselves short. I did things way beyond what I ever dreamed of doing.

Overleaf: Portrait of the combined STS-84 and double Mir crews for changeover, showing 10 people in space. This equalled the largest number of people in space at any one time, a record which was set on a previous Shuttle-Mir flight.

THE COLLISION

MICHAEL FOALE: Approaching Mir from a distance, it was the biggest structure I'd ever seen in space. As you enter the space station, the noise changes, and then you get the smell as you pass from the shuttle, which is all clean, into Mir, which smells kind of musty and oily, a bit like a garage or a machine shop. I was trying not to put my hands on the walls, because you can knock stuff off, but you've got to somehow propel yourself through, so I was just pulling very gently with my fingertips. It was then that I noticed that along this trail a strip of Soviet-red material stretched all the way to the node. I think Charlie[10] then asked Vasily what the red material was for, and Vasily said, 'I wanted to make a proper welcome for you, so I laid down a red carpet.'

I did my handover with Jerry over the next two or three days, and just before he left, Jerry said, 'There's one thing I haven't told you, but I need to.' I was pretty saturated with information at this point. He said, 'Vasily and Sasha set up this remote-control system to fly a cargo ship called Progress.' I knew that these were cargo ships that brought up critical supplies, and then they stayed attached at the back end of Mir. After two or three months, they were filled with trash, so they were pushed away by springs and the last bit of rocket fuel they had on board was fired, and they burned up in the atmosphere.

Jerry said, 'They were told by Moscow to manually control the Progress to dock back with Mir.' I said, 'Why would they do that? Does NASA know about this?' Jerry said, 'No, I don't think so.' The Russians had decided many years prior that a radar docking system was an important tickbox to have, because when they were developing their first space stations, they knew they needed to know the speed and distance to bring spaceships in together, otherwise you might crash or miss one another. Jerry then explained that he had witnessed an experiment to dock a cargo ship without using the radar that tells the automatic docking system the distance and speed, and that it had resulted in a very near miss.

After the shuttle left [taking Jerry with it], it was really quite pleasant and calm, and my fears about not being accepted, or not being well understood, went away. The next day, Sasha floated in and said, 'Michael, how are you doing?' I said, 'I'm finding myself. Thanks, Sasha.' And then he said, 'Come and have some tea.' I said, 'Really?' And he said, 'Yeah, come on, have some tea. You can do that a bit later.' So, I floated after him, and we just chilled with Vasily and chatted for about maybe twenty or thirty minutes as a tea break.

Looking back, I think Jerry was a polite guest and they treated him as such. Although he still integrated with them just fine – he demonstrated that in the fire – my approach was more gregarious. Because I come from the UK, I also had a transatlantic viewpoint of the cultures, and I believe I was more comfortable working with the Russians than the Americans were. And let's be honest: the Russians like tea, and the British like tea. We all understood that we needed to fit into the Mir programme and with the crews on board who didn't speak English, so I would invite Vasily and Sasha to come and watch movies with me. I'd chosen *Total Recall*, *2001: A Space Odyssey*, *The Right Stuff* – there were a number of dramas – and NASA had dutifully gotten the copyrights and put them on tape so I could watch them. I had to translate them all, which really helped my language skills. Vasily would always float in and hand me $1 and say, 'This is my fee for the movie.' I'd put it in my pocket, and we'd watch it, and he almost always fell asleep. The next morning, I would take the dollar and put it back on the wall, and he'd do that again and again.

---

[10] Charlie Precourt, commander of the Shuttle mission STS-84 which delivered Michael Foale to the Mir space station.

MICHAEL FOALE: I had been there about six weeks, doing my experiments, when I noticed that Vasily and Sasha were setting up the system to fly in a cargo ship. I asked them what they were doing and one of them said, 'Well, you know that cargo ship we just let go full of our trash? We're going to try and fly it back in.' I remembered what Jerry had warned me about, so I started asking questions. One of them said, 'You don't need to know much about it, Mike. Just do your experiments.'

The next morning, they had it all set up so I went to the back end of the station to where I should have been able to see Progress coming in, but I couldn't see it. I floated back and said, 'Hey, Vasily, I can't see it.' I could see he was completely absorbed. Sasha was looking down through a window by his feet, and as I floated up, I noticed on the TV screen that everything was wrong. The orientation of Progress coming into the station was out by about 90 degrees.

Sasha said, 'Michael, go to the spaceship!' He meant the Soyuz, because it was the lifeboat. The extreme urgency in his voice made me think, *'He's serious,'* so I didn't ask questions. It was one of those emergency moments.

SASHA LAZUTKIN: There was a large window aperture in the floor. I was standing next to Vasily at that time, and I saw the ship sweeping by beneath me. For some reason I compared what I saw with a shark. And then, it hit the station. The collision was heavy and the station rocked.

After a matter of several seconds, perhaps three or four seconds, the emergency siren wailed. The corresponding display had lit up, and Vasily said, 'Depressurisation'. That meant one thing: I had to work to eradicate that depressurisation.

MICHAEL FOALE: As I heard the crunch of the collision, I thought, *'Crikey, it's hit us.'* Because the aluminium is only about three millimetres thick, I started to look at the bolts [where panels were joined together] and thought, *'If it's going to break it apart, it's going to break apart here.'* My next thought, because I'm a diver, was, *'Breathe out.'* I didn't want to burst my lungs when depressurisation happened – although this would have been futile if there was no air to breathe afterwards – but it didn't happen. I stopped looking at the bolts and thought I'd better get on with the Soyuz when I felt my ears pop [an indication that the hull had been breached and air was escaping].

Mir was losing its atmosphere, a mixture of oxygen and nitrogen, just like we have on Earth. And that leak and the pressure drop is what was causing a whistling sound – it was air escaping out and going into space.

LUDMILLA LAZUTKIN: We had gone to the seaside in Tunisia for a vacation, me, my parents and children. You could even see the station flying from there – its route passed exactly over the place we were staying – which I was thrilled with because you can't see the station in Moscow. I just wanted to walk around and tell everyone, 'My husband is flying over us right now, he can see us.'

We even had a conversation [with Sasha] once, we called Mission Control Centre (MCC) from the hotel and they let us talk to him for just a minute. When the vacation was over, we landed back in Moscow and were met by the driver, who said 'Oh, by the way, did you hear that there was a collision?' I said, 'What?' He said, 'Well, they said it was some kind of depressurisation.' That's all.

On the news, they just kept repeating that there had been a collision, but no one knew anything. Any ignorant person understands what depressurisation is. You know that, in the vacuum of space, depressurisation is death. So we sat and waited. The waiting hours are probably the most intolerable thing in any situation like this, when you don't know what's going on with your loved ones. We were like zombies... we just prayed and hoped that everything would all end well.

SASHA LAZUTKIN: I started doing what I had been trained to do. I flew to the spacecraft, prepared it for immediate abandonment of the station, because in the event, we would need to get urgently into the spacecraft. I understood that my next step would be to search for the location where the hole had been made. I saw which module had been struck and flew into it, where I heard the fans working and heard an extraneous new noise: that was air escaping from the station.

I realised that that was the site of the depressurisation and that we had to isolate it from the rest of the station. I flew off and reported to Vasily that that module had been damaged. Meanwhile, he had begun establishing communication with Earth, while continuing to monitor the air pressure. All the time he was saying, 'The air pressure is this...760mm...750...740...' the pressure was dropping. And he was reporting that back to Earth, that the docking had not taken place, that there had been a collision, the station had been depressurised and we were working to tackle that emergency situation.

After some difficulty, Michael and I managed to close off the module, while Vasily was busy preparing a system that would release air for us. The pressure stopped falling and we reported back to Earth. However, the collision, when the vessel hit us, had caused the station to spin; to tumble. That meant the solar batteries stopped receiving sunlight, and in the end, we ran out of power.

Below: Russian experts in the Mission Control Centre discuss ways to stabilise Mir and stop it tumbling out of control. Vasily would later explain that his actions averted a much more serious catastophe in which they all could have died.

MICHAEL FOALE: We were always moving eastward, so we were about to lose radio communication with Moscow and go into darkness over the Pacific. Vasily came off the headset, and he said almost nothing to us. He was in emotional shock, already thinking he was guilty and that he was the one they were going to blame, and that it was a calamity equal to the worst destruction in the Soviet Union, something like Chernobyl or something, because Mir was the pride and joy of the Russian space programme.

As we went into darkness, another alarm went off and [a warning light] lit up: low voltage. First, we heard the fans stop blowing air around us, and then the lights in the other two modules that were still connected to us shut down. All you could hear after a while was a kind of click, click, click as the station cooled down and contracted in the cold of space. Other than that, it was absolute silence. It was a dying station.

SASHA LAZUTKIN: When we fly into space, it is never quiet on the station: the engines are working, the fans, all the time. There is always this noise, much like being on an aeroplane: you can sit and chat but there is a constant noise. It was just the same on the station.

However, when the power disappeared, there was this silence, and I remember that my ears even hurt from it. I covered them and thought, *Now I have heard cosmic silence*. It was absolute silence. In that cosmic silence, when nothing is working, there was no light at all, and on the Earth beneath us there was not a single city light because we were over the ocean.

But the stars! An incredible number of stars! We were flying round the southern part of Australia and not far from there was Antarctica. And over all Antarctica, we could see the aurora polaris, from 400 kilometres above Earth. The flames were ghostlike, ever so slightly light blue, slightly green, slightly pink, slightly yellow. There was a sensation that the flames were reaching up for us. They were all across Antarctica, an enormous expanse: it was spellbinding and we simply froze.

LIVING IN SPACE 159

MICHAEL FOALE: The sunrise was coming in twenty or thirty minutes, so we looked outside to see what was happening to the station. I noticed the stars were now moving past us. Mir doesn't usually rotate relative to the stars, but I could see they were moving, which meant we were spinning. I tried to estimate how fast we were moving, so I went to another module, and looked out the window and put my thumb out against the sky. Sasha and I knew from schoolboy physics that when held at arm's length in angular measure, the width of a thumb was approximately one degree. We used this to time how long the stars would pass our thumbs, and work out the speed of our rotation.

I said to Vasily, 'We're tumbling at one and half degrees per second.' From my training I remembered that, about 13 years ago, a cosmonaut had reorientated Mir towards the sun using the Soyuz. Eventually, Sasha and I convinced Vasily to go to the Soyuz and try to turn it on.

He came back with a really ashen face and said, 'I can't turn the Soyuz on.' In hindsight, this was a big design flaw in the Mir construction. There was no power in Mir, so there was no power in the Soyuz, and you couldn't separate the vehicles. You couldn't even turn on the Soyuz. It was bad, but I didn't feel scared or worried. I was just thinking, *This is a frustrating technical problem we have to solve.* So, we waited and went into darkness one more time.

The saving grace of this story is that the Russians never got rid of the back-up mode of operating the Soyuz, so the individual rocket thrusters could be fired individually without the computer. Sasha told Vasily to use the manual mode. It was nighttime now, so we were using the stars to tell if we'd changed the spin or not, and Vasily basically followed our instructions. [In this way, we were able to get the station back into] the proper orientation, and as we came into sunlight, the solar panels were pointing towards the sun, and the power returned. The lights came on, and the fans started up, and we'd pretty much got it solved.

I looked at Sasha and we smiled. This was probably our first smile since the collision. Vasily didn't smile, but he was more relaxed, and it was sort of a collective sigh of relief. As I reflected on what had happened later, I was just thinking about Vasily and the blame that was going to be aimed at him. I went to bed feeling sad about that, but I also felt pleased with what we'd done to stabilise Mir. I went to sleep a very happy camper.

Right: The Mir space station and the Moon captured by one of the STS-91 crew members aboard the Space Shuttle Discovery.

SASHA LAZUTKIN: After that, the normal routine picked up again. We had succeeded, but we were not euphoric. We simply gave a sigh and carried on with our work. There just wasn't time: communication was restored with Earth and we started to measure everything. Down on Earth, they started working on stabilising the status. We still had to assess what had happened.

It was Earth that started to assess it, determine why there had been that collision, what had happened. We found out just how powerful the impact of the collision had been. The freight vessel weighed seven tons, and it struck us at a speed of three metres a second. That is fast.

All of those events that unfolded, they somehow managed to desensitise the soul. It was only after all of this that I thought of myself as a doctor; I sensed the station with all my body. I knew what had broken, where something had malfunctioned. I just acquired that confidence because I fixed everything with my own hands.

When I first arrived, I was afraid of the station. I remember the first operations I performed, pressing buttons. They came with the shivers: what if I was doing the wrong thing? Later, no problem, I fixed one thing, then another; it made no difference to me anymore. I could touch space. That is how I perceived it. For me, that thing, that big structure, inanimate, it was a living organism, a body.

I still get the question here on Earth: who should explore space? Robots or humans? Everyone says that it should be robots because humans are very delicate and need the creature comforts, but robots don't. The robots will do the studying. But even in that mission, I realised that a human is the strongest link in that chain of design of space equipment, the study of outer space and so on. Any device, any robot, can break down. But a human can always find a way to fix things.

Here, once we had fixed everything, sure, we were tired from all that, but I had this feeling of being a strong person. I acquired the understanding that I can fix all of it. It is the human who conquers space, not robots.

RHONDA FOALE: I was on a trip to Kentucky to visit my relatives, so I was at my parents' house. The phone rang at like six in the morning, and my mom answered it. It was Terry Taddeo, the flight surgeon who had been in Russia with us and was assigned to Mike to monitor him during his mission. Terry said to me, 'Rhonda, Mike is safe. Everything's OK now, but an accident has happened.' And then he told me a few of the details. At the time, I didn't really realise fully what had happened, or [know anything about] the ongoing threats and failures that kept happening after that. I didn't find out until a couple of years later, and then I thought, *I'm so glad I didn't know this at the time.*

Right: Damage to Mir's Spektr module and its solar arrays, caused by the Progress supply ship collision.

Alexandr Lazutkin and Vasily Tsibliyev spent a total of 187 days in space, returning home on 14 August 1997. Michael Foale returned to Earth 53 days later.

MICHAEL FOALE: As Sasha and Vasily were leaving in the Soyuz, I tried to anticipate what was going to happen to them when they arrived back in Star City. I knew that what Vasily and Sasha had done was save the space station Mir. Their persistence and desire to make it right, to try and undo the collision's consequences, was heroic. But I knew the reception might not be good and I was worried and anxious on their behalf. At some point when Sasha and Vasily were in Star City after they landed, a manager said, 'Vasily, you were the one responsible for this disaster,' because management tends to blame the Russian employees in Russia. Apparently, Vasily said, 'Yes, I'm the one who should be blamed. I'm the commander. I'm responsible.' As I understand it, Sasha, who was always utterly loyal to management, spoke up and said, 'No, I don't think it was Vasily's fault. It was someone in management's fault for making the decision to not use the radar.'

As a result [of the collision], Vasily was not assigned to a flight again – he was given a minor role managing training at Star City – and Sasha wasn't assigned to another flight for many years. After two or three years, because he admitted guilt and was apologetic to senior management in Moscow, Vasily was promoted to a two-star general and became head of Star City. And Sasha was still at the same low level. Only much later on was Sasha assigned to an International Space Station flight.

SASHA LAZUTKIN: Vasily was the commander of the crew. He understood that when we returned to Earth, he would be the first one to be blamed. He spoke very calmly and explained that he was unable to stop the vessel, so it struck that module and the station was depressurised. We had taken the requisite actions. His report was clear and consistent: there was an order to events, what had happened, why it had happened and what we did.

The thing is, despite what had actually happened on the station, people would say, 'Look, things have never broken before like they did on this mission. The station has been flying for ten years in fantastic order. One crew goes up and it all starts to fall apart. We need to work out why.' These statements appeared in the media and were made by people in influential positions. We were under this unspoken suspicion from the start, we could sense it when we were talking to mission control. We could already sense that the top brass didn't trust us.

I remember that back when we were still training on Earth, even experienced cosmonauts and specialists said, '[docking] is a boring process. Nothing can happen there.' Therefore, it was a total shock when that accident happened. And so we had to make sense of it together with the specialists at mission control, using the information that we had.

LUDMILLA LAZUTKIN: The press harassment began when President Yeltsin was asked, 'What's going on at the station, Mr President? How can this be happening?' And he said, 'It wasn't the equipment that failed. It was human error.' There had been no inquests at that time, so it was not yet clear what had happened, but he needed to show the Americans that it was all fine. Of course, the race for interviews began. This period was very difficult, because the journalists waited outside our house and my work, so I moved to the dacha. It was easier for Larisa in Star City, because it was closed off.

The judgement was passed right away. That's what was despicable, both on the part of the President and on the part of the journalists. Later, it was confirmed that the committee, which spent six months working out what happened, did not file a single complaint against the crew. Did anyone apologise? No. Yeltsin seemed to admit his mistake and put the Hero of Russia Gold Star on my husband's chest, but even this was done in a very behind-the-scenes way. It's usually done in a celebratory manner, whereas they just called him to come in on 13 April and presented the award with only two or three people present, and no journalists present. This was also the only time where there wasn't even an official photo taken.

I think it was very difficult for [the children] to live through all this. One was a teenager; the other one was very young. My biggest mistake was not explaining anything to the little one. What did she pick up and how did she make sense of it? She didn't ask questions. She might have overheard some things, and some things went over her head, but she saw that her mother was worried and suffering.

ZHENYA LAZUTKIN: I heard all those strange words like 'depressurisation', 'non-standard situations', and so on. Being a child, I didn't understand the meaning of them, but those words had a certain colour to them and an atmosphere in which they were spoken. And so, as a child, I could only process them in my inner world, feel them, understand what they were for me.

I do remember knowing that I shouldn't show Mum that I was worried, because she was worried enough as it was, and why should I add to that? So, even if I'd heard something, I would make it seem as though I hadn't. I was terrified, but I didn't show it. When you don't know what is happening, you are in a continuous state of nervous tension, and that is really hard.

Right: Sasha Lazutkin on Mir.

LUDMILLA LAZUTKIN: When I saw Sasha's legs [the night after the landing], they were bones covered in skin. It was a shock for me, because I had sent a gymnast into space – you can imagine the muscles, the biceps, the abs – and he came back completely weak. Moreover, after the flight, he could never reach that level of fitness again.

When we arrived in Star City, his doctor helped him get to his room, and a lot of people came to congratulate him on finishing the flight. Alexei Leonov came, which we were extremely grateful to him for, because there were a lot of people who tried to distance themselves, thinking *'Maybe they were at fault, so it's better not to shake hands with them'*. When you feel supported by a person like Leonov, it gives you the energy to continue fighting for your honour and dignity.

When I put Sasha to bed and he stretched out, he was enjoying the fact that he was finally lying horizontally. Suddenly, my laconic husband started to talk. He spent half the night talking. It felt as if he was talking to put things on record. I was afraid to breathe out or say anything because I'd never heard him talk so much, and I had the feeling that he was saying it out loud to try and process it for himself. It was like a confession, such a sincere, emotional speech. I had never heard him talk like this before or since. I was dumbfounded. First, by the events [he described], and, secondly, by the emotions that were coming from him. This was, of course, an incredible experience. I thought, *'God, I just want to stay awake and listen to it all right till the end.'*

After her dad returned from the flight and went through the rehabilitation, I began to notice that Zhenya was looking pale. She felt lethargic, weak and tired all the time. This was very unusual for an eight- or nine-year-old child. The doctors said, 'It may be fatigue, or perhaps overexertion.' They came up with all sorts of diagnoses. Then, in the six months from the winter of 1997 until the spring of 1998, she stopped walking and lost the use of her legs. She was practically not alive anymore. The doctors told us, 'We don't know what this is. It could be a blood disorder,' and used all sorts of scary words – the worst things you could think of.

We were advised to see a psychoanalyst, and he started to pull her out of it. He said that it was the stress of her dad's flight and everything that she'd lived through with us that did this. It had seemed to us that she didn't realise what went on, and that it all went over her head, but she had internalised it. It took us two years to get her back on her feet. From the point of view of our family, I believe that Sasha's flight was actually a turning point in the relationships between us all. This nightmare that we had gone through brought us closer together as a family.

Below: Russian cosmonauts Vasily Tsibliyev and Sasha Lazutkin on their return to Earth.

ZHENYA LAZUTKIN: Dad told me that when you fly up into space and look back to Earth you can see it as a whole. There are no lines, like those drawn on a map. Up there, you come to realise that we are all one family living on a single planet, and it's really not that big, compared to the scale of space. I think your view of life changes, and the notion of borders falls away.

I recently attended one of Dad's talks, and I keep thinking about this thing he said: that while we only look down, we'll have conflicts and discord. But if we look up, then the opposite will occur, and we will all come together. I really liked that thought that if you are looking up, it doesn't matter how much territory you have beneath your feet – you will know that the world is enormous, and we are one family.

THE LEGACY OF THE SHUTTLE-MIR PROGRAMME

MICHAEL FOALE: The ISS came out of the Shuttle-Mir programme. Not haphazardly, not randomly: it was the end goal. That's what President Clinton and Boris Yeltsin had in their minds when they told NASA [to] go work with Russia.

In that first stage, we learned to work with each other. We learned to respect each other. We learned to rely on each other. We learned each other's cultures and languages. Today, the legacy out of that, as we continue to operate the space station with Russian and American partnership, is knowledge, and the experiences of thousands of engineers, scientists, managers, support workers and families, both in the USA and in Russia, having interacted.

The stakes were really high: we knew that the ISS wouldn't happen if Shuttle-Mir was unsuccessful. If I had left because I thought it was too much to put up with, or for political reasons, the consequences would have been dire to the programme, to cooperation with Russia in the future and to building the International Space Station. I didn't feel that was all on my shoulders, but I was aware of the consequences. In the US Congress, the fire and the collision both raised the issue of whether Phase One should be cancelled – it was not a popular programme and anything we had done on Mir could have swung the votes in Congress as to whether to continue it. It would have been shameful for me to leave Mir, and cause the demise of Phase One, and probably the demise of Phase Two, the ISS. We would have delayed in having an International Space Station, and it probably would have been without the Russians.

People sometimes ask, what did the US get out of this cooperation in the Shuttle-Mir programme? And what did the Russians get? The immediately obvious thing is the knowledge of how the other side does things. It's holistic in that if you work together, you get a bigger product than if you just had the separate pieces. We talk a lot about the astronauts and cosmonauts, but supporting us and working towards the same goal are all these other people who have dreams and aspirations for peace and adventure and exploration, in both America and Russia. And we must not forget Europe, which was, in many ways, the glue [that brought us together]. This multinational cooperation is the legacy.

Left: Michael Foale with his family meeting President Bill Clinton, in the Oval Office.

JERRY LINENGER: I love being a trailblazer, and I think my life was worth something. I'm proud of my actions and think I did the job to the best of my ability. I also think we should all strive in a small way to advance humankind. You can do it on an individual level. If you're in a position of power, you can do it and really affect things. Or if you're privileged enough to be an astronaut, you can maybe get humanity looking towards common goals and something bigger than ourselves. It was an honour and a privilege for me to represent the United States of America, my country, and interact with the Russians, our Cold War enemies, to find a common goal and hopefully help our spirit of cooperation here on Earth.

The legacy of Shuttle-Mir is that we stuck together and showed it could be done. It was rocky at times, but it undoubtedly led to international cooperation on the International Space Station.

SASHA LAZUTKIN: Michael came into my life back then and has remained in my heart as someone who is sunny, open, kind and warm. We don't meet that often, very seldom in fact. But I'm always comforted by the knowledge that Michael lives on this planet. I know him, he knows me, and this makes me feel very good.

You know, when the shuttle crew arrived at the station, and we were all sitting at one table, we were chatting about this and that, and I felt that out in space all people are equal. They are not the representatives of America, Russia, France, or some other country. They are people of the Earth.

The moment we are back down on Earth, we return to representing a particular country. Politics begins to rule over everyone. We can fight in wars against each other, we can experience joy with each other, we can work together or refuse to cooperate. Politics somehow directs us down these channels.

This is a given; like it or not, it's how it is. But the moment we depart from Earth, politics no longer plays a part up there. It vanishes from our conversations, from our relationships, from everything. It's just not there. Space does not tear us apart; it brings us together.

Above: Jerry Linenger performs checks on his Russian pressure suit.

Overleaf: Jerry Linenger during extravehicular activity (EVA) outside Mir space station, 1997.

LIVING IN SPACE

PART THREE:
THE NEW SPACERS

'We were Apollo's children. We were promised [that] the sky is no longer the limit.'

CARLOS GARDELLINI,
Entrepreneur

INTRODUCTION

The Shuttle-Mir experiment was a success. There had been a few scary moments, but the inventiveness and ingenuity of both astronauts and cosmonauts, working together, meant that nobody had died, nobody had quit, and nobody from either side seemed to regret the decision to work with each other. While cooperation between America and Russia remained elusive on Earth, in space at least, it was achievable. NASA had learned much from the Russian space agency about adapting to long-term space habitation. The Russians too had learned from the Americans, and at the same time had also managed to keep their space programme alive with the help of US dollars.

Shuttle-Mir was always intended to be the first stage of a more ambitious plan. The end goal was to build an international space station. The ISS was conceived as a modern, state-of-the-art, permanently inhabited orbital space laboratory, built and operated jointly by America and Russia, and thirteen additional international partners, a physical symbol of a unified, peaceful species learning to live together beyond their home planet. But the shine of cooperation quickly faded as Mir itself became a burden. With the first modules of the ISS now in orbit, Mir became a distraction for NASA and they wanted it gone. Designed to be operational for just five years, by the turn of the century, Mir had been in orbit for nearly 15 years. It would take huge amounts of money and a massive commitment to repair the leaking space station and keep it operational. NASA, knowing that money was still tight for the Russian space agency, decided the best course of action was for the Russians to deorbit Mir and focus all their attention and resources on the ISS. The prospect of deorbiting Mir was difficult for Russia to accept, as deorbiting really meant nothing more than a controlled re-entry, essentially burning it up in the atmosphere and crashing any remaining fragments into the

Previous page: A portion of the SpaceX Cargo Dragon vehicle is pictured lower left docked to the ISS, orbiting 264 miles above northern France.

Pacific Ocean. The prospect of the pride of the Soviet Union, the symbol of Soviet greatness, being dumped into the sea was not an easy pill for any Russian to swallow, no matter how logical it may have been. But with little leverage against the United States, there appeared to be no other option. That was until a group of American anarcho-capitalists arrived in Moscow, with a plan to buy Mir and keep it in orbit.

This band of entrepreneurs represented a new character archetype entering the story of space – the private space developers. Up until this point, human space flight had entirely been the domain of governments, as who else besides nation states could afford to access space? There was a growing frustration, however, from a community of space enthusiasts about the progress governments were making in delivering what they believed was the promise of the Space Age: access to space for everyone. The 'new spacers', as this community was known, comprised space fanatics, visionaries, radical thinkers and the occasional billionaire. Growing up in the wake of the Apollo Moon landings, they bonded over a frustration that a future they felt they had been promised – comprising of flying cars, jet packs, and the ability to routinely visit other worlds – had not materialised. When Walt Anderson, a self-proclaimed 'new spacer', heard about the precarious situation with Mir, he came up with the radical idea to buy it. This was the start of the commercial space venture, and the reluctant beginning of government agencies ceding control of space to private enterprise. It was going to be a difficult but necessary transition, as, for the first time, questions such as who controls space and who is allowed to go began to be asked.

Opposite: The SpaceX Falcon 9 commercial rocket blasts off carrying the Dragon capsule on a resupply mission to the ISS on 28 June 2015 in Cape Canaveral, Florida.

ORPHANS OF APOLLO

Below: Buzz Aldrin on the Moon in a photograph taken by Neil Armstrong, who can be seen in the visor reflection along with the Lunar Module Eagle, and the US flag.

Right: A child watches as the first televised pictures of the Moon's surface are beamed back to Earth in detail. The Ranger 9 unmanned probe shows Americans an unprecedented view of the surface.

After the Soviet Union launched the first artificial satellite Sputnik and the first space race began, huge progress was made in a short period of time, with the first Moon landing happening a little over a decade later. This incredible technological advancement was inspiring to people beyond those who would one day go on to fly in space as part of a national space programme. Carlos Gardellini, whose childhood coincided with the Apollo programme in the USA, was one such person who saw his future in space as being almost inevitable as he was growing up.

CARLOS GARDELLINI: 20 July 1969 – that's the pinnacle of human achievement. That's something that, 55 years later, we haven't been able to duplicate. Neil [Armstrong] puts that footprint on another body, and that changed all of us. That's also the moment I knew I wanted to go to space.

I knew if I got good grades, which I did – straight As – and if I worked hard, which I did – other than ditching school every time something in space happened – I'd be an astronaut, guaranteed. From my perspective, that was the natural progression of things: 1969 Moon, 1979 Mars, 2001 colonies everywhere. How could it not happen that way? [President Kennedy] said in 1962, 'Before the decade is out,' and eight and a half years later, these guys were prancing on the Moon.

178 THE NEW SPACERS

In the 1960s, Walt Anderson, who later became an investor in private space initiatives, also believed the 1969 Moon landing promised a spacefaring future.

WALT ANDERSON: When I was five or six years old, I was very interested in astronomy and the structure of the planets, stars and galaxies. But I was also a reader of science fiction at a very young age. So, that combination of science and science fiction piqued my interest in space. And I really believed that it was possible to go to space when I was very young.

I was in elementary school at the time of the first Moon landing. It seemed like progress, a step for mankind. I really believed that humans needed to go out into the universe as part of our destiny.

Rick Tumlinson, who became an advocate for the involvement of ordinary people in space exploration, was inspired by the technological advancements of the Apollo space programme but also by the science fiction of the time

RICK TUMLINSON: My father was a sergeant in the army who was assigned to a base in the UK, and we lived in a council flat in Peterborough at the time. I remember sneaking downstairs to see this new show called *Star Trek*, and, right around the same time, seeing the first steps on the Moon. It was fuzzy, but I got that it was important. But I was more excited by *Star Trek*, because it was cooler.

But it wasn't *Star Trek* by itself. It was that *Star Trek* was adjacent to the Apollo programme. So, you were seeing what was happening right now, and then you were seeing this depiction of where it could go. And in a child's mind, they came together, and I thought, *By the time I grow up, I'm going to be on the Enterprise and heading out into space.*

I believed at that point that we were just going to keep rolling. There would be cities on the Moon and then we would go off to Mars. But it didn't happen, so, later on, I got disillusioned, and then I got angry. And then I got active and started trying to make it happen.

Richard Garriott, the son of NASA astronaut Owen Garriott and who would one day fly in space himself, was brought up to believe that rapid space development was inevitable.

RICHARD GARRIOTT: I had a pretty unique childhood. My father was a NASA astronaut, originally hired to be part of the Apollo missions to the Moon. My mother was a professional artist and a naturalist. They were both educators, so I had a really strong STEM, or STEAM if you include the arts, background growing up, which led me into my initial career, which was video games.

There was an artist by the name of Bob McCall who worked with NASA to envision the future. He was always producing paintings of vehicles in space, or domed cities on the Moon, or big space stations. I had complete confidence that this was what was going to happen next, so it was actually quite surprising to me when the Bob McCall version of the future didn't happen in five years, as I assumed it would. Turns out it's a lot more expensive and difficult than we thought.

The Apollo programme had an impact on those outside of America too, including on Anousheh Ansari, who was a young girl in Iran when the first Moon landing occurred.

ANOUSHEH ANSARI: The Moon landing impacted everyone. It sort of created space dreamers all over the globe, and it did in Iran too. I loved *Star Trek* growing up, and I watched thinking that one day there would be a Starfleet Academy and I would apply, ready to go on the Starship Enterprise. I think this promise of everyone's going to go to the Moon, and we'll have a hotel there, and the rise of science fiction and how space will be part of humanity, started in that [Apollo] era.

Later on, I found that, no, things didn't progress as we'd hoped after we went to the Moon. But I never thought going to space was a crazy dream that wouldn't be possible. I knew it would be difficult, but I was ready to meet any challenge to do it. I believed in it so much.

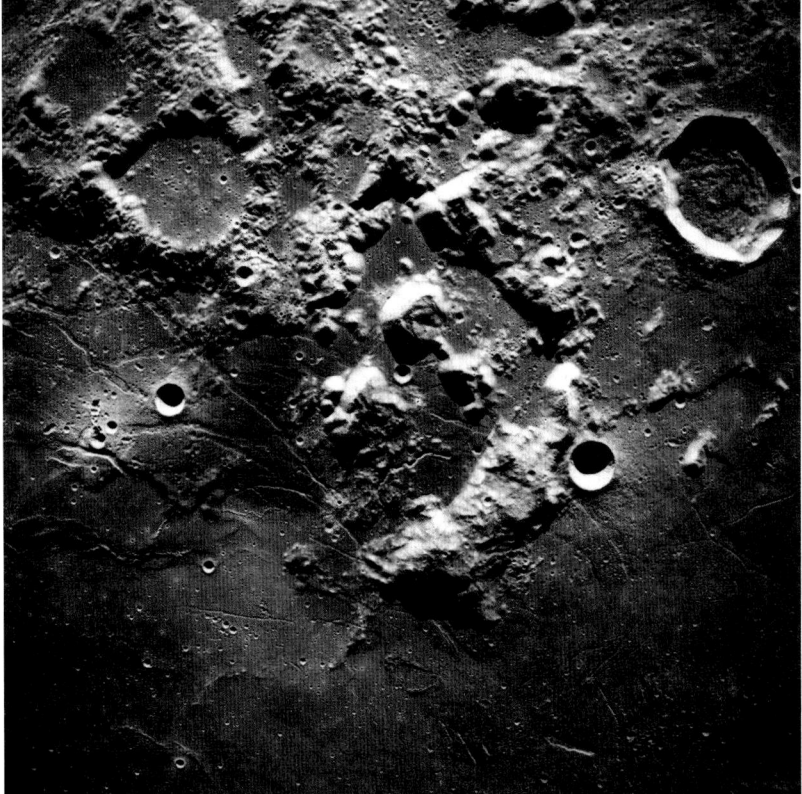

Above: Anousheh Ansari as a child in the 1970s.

Below: The NASA Apollo 17 landing site.

Left top: The Garriott family.

Left: Robert McCall's 1997 mural Accepting the Challenge of Flight at NASA's Dryden Flight Research Center, is focused on portraits of actual Dryden employees. Flight research aircraft of that era fly above, and his ever-optimistic view of the final frontier is in view at the top.

CARLOS GARDELLINI: 11 December 1972 is the day the music stopped – the last lunar landing. That's when, as a nation, we decided that we'd shift from looking outwards to looking inwards. No more Moon missions [with astronauts], certainly no Mars missions since the Viking missions to Mars followed Apollo quite swiftly. If the Earth were an apple [in terms of human space flight], the Earth's atmosphere is the thickness of the skin. For the next 50-plus years, we're going to devote ourselves to barely piercing the skin of the apple. Politics always wins. Apollo 18, 19 and 20 were already built, fully tested and ready to go, [but they were] cancelled, thrown away. If that isn't the final blow, that we go from looking out to turning inwards, I don't know what is.

When you're living in it, you don't realise that this is the moment: this is when it all goes downhill. You're just living your life. It's only after enough time passes that you realise, 'Oh, that was a moment back then.'

We were Apollo's children. We were the first generation after the Greatest Generation,[1] and we were promised [that] the sky is no longer the limit. This is science fact. And then it didn't happen, so we were orphaned.

I always saw getting into space as my goal. And for me, the shortest, shortest distance there was studying to be a rocket scientist. So, get your engineering degree, get a master's, move to Houston … astronaut. Of course, you do the incidental stuff, too – get your pilot's licence, scuba certified. I'm ready, man! Where's my jetpack? Where's my flying car?

I graduated and got seven job offers – if you were a rocket scientist at this time, you were golden. What sold the deal at Hughes [Aircraft] was that they took me to the production line in El Segundo, California. I remember walking in with my jaw to the ground.

There are only two kinds of rocket scientists: those who get their stuff into orbit and those that ain't worth shit. Within the first month, someone called Art Zapz pulled me aside and said, 'I need this quick design. Can you do it?' Long story short, the space shuttle had deployed two huge satellites, but the motors had failed on both of these birds, and they were going to re-enter and burn up. The rocket scientists at Hughes came up with this idea: 'The shuttle's going back up there in November? What if we capture these things and bring them back down to Earth?' I started work in March 1984 at 23 years old with the promise that I was going to get my hardware up in November of that year. It was madness. We brought the satellites back down and refurbished them. The next year, they went up into orbit and worked flawlessly. I was riding high, so I immediately applied to the astronaut corps. *How could they not pick me? I've already got my hardware up there!* Back then they had this programme called 'payload specialist'. It was a way to cut the line. Who better than me to deploy our satellites? Of course, I was rejected, and [my colleague] Greg Jarvis was picked. No one knew our birds better than Greg.

I was in bed at 8.39am on 28 January 1986, and my mom came rushing in. I was sick, so I'd already decided I was going to stay home that day. Challenger blew up at 11.39am in Florida, which is 8.39am in California. At 8.40am, I called Hughes: 'Is Greg all right?' 'No, he's not. He's dead.' Challenger personalised it in ways that I could never properly explain.

[Challenger] made me redouble my efforts to [get into space]. You get knocked down, you get back up. There was no hesitation for me. Proof of

---

[1] The term 'Greatest Generation' generally refers to Americans borns roughly between 1901 and 1927. This generation came of age during the Great Depression and then served in World War II, either in the military or on the home front.

Below: The NASA Apollo 17 crew in front of the Saturn rocket (left to right): Harrison Schmidt (Lunar Module Pilot), Ron Evans (Command Module Pilot) and Gene Cernan (Commander) seated on the Lunar Rover. Apollo 17 was the final crewed mission to the Moon.

the pudding is in the eating: I moved to Houston in 1988, two years after the Challenger accident. I lived at 1713 NASA Road One, across the street from the space centre. I was hanging out with all the best NASA designers and with astronauts at the Outpost [Tavern], where they go to drink beer and bitch about being an astronaut. Five years I gave it a shot in Houston. I hung out for two full selections and didn't make either of them. So, I applied to the astronaut corps three times. Once for the payload specialist position, which was cutting the line, getting in front of all the other astronaut candidates, and then I moved to Houston and applied two more times and failed there too.

Below: Astronaut Harrison Schmitt and the Apollo 17 Lunar Roving Vehicle (LRV) photographed behind 'Tracy's Rock' in the Taurus-Littrow Valley, on the final EVA of the final Apollo flight to the Moon.

NEW SPACE

Physicist and author Gerard K. O'Neill was a big influence on the people who felt left behind by the mainstream space industry.

RICK TUMLINSON: Reading *The High Frontier* by Gerard K. O'Neill was probably the most pivotal moment in my life in terms of moving from, *This is cool, I hope it happens, but I don't know how to get engaged in it,* to, *This is going to be my life.* O'Neill was a Princeton physicist who was involved in the development of particle accelerators like CERN. In his Princeton physics class in the early 1970s, he posed the question to his students, 'Is the surface of the planet the best place for civilisation?' And the answer he received was no. He embraced it and wrote *The High Frontier*.

At that point, anything to do with space was government and academic only. Regular people were not included. Part of the book is written as a letter from a little girl growing up in a space colony. You couldn't be less 'Right Stuff' than that. This made people like me think, *I can be part of this, even though I'm not an astronaut candidate or a physicist.* There was this high-school kid in Florida named Jeff Bezos who had a book club featuring *The High Frontier*. It changed his life. The book basically gave permission to dream to an entire generation that had been locked out of the aerospace industrial complex of the Apollo programme.

When O'Neill wrote *The High Frontier* and got us all excited about this possible future of opening up that frontier, all of it was based on a lie that was being fed to us by the aerospace industrial complex that there was this thing coming called the space shuttle, and it was going to fly 50 times a year, and it was going to bring the cost of going to space down. If that's your premise, then economically speaking, it becomes possible to talk about building a human community in space. But as we started rolling into the 1980s, we began to realise it was a lie. The space shuttle established a price of roughly $10,000 a pound for carrying payloads to space. It didn't open up anything. I know people who flew on the shuttle. They're amazing people. It was a magnificent machine. But it was not what we were sold.

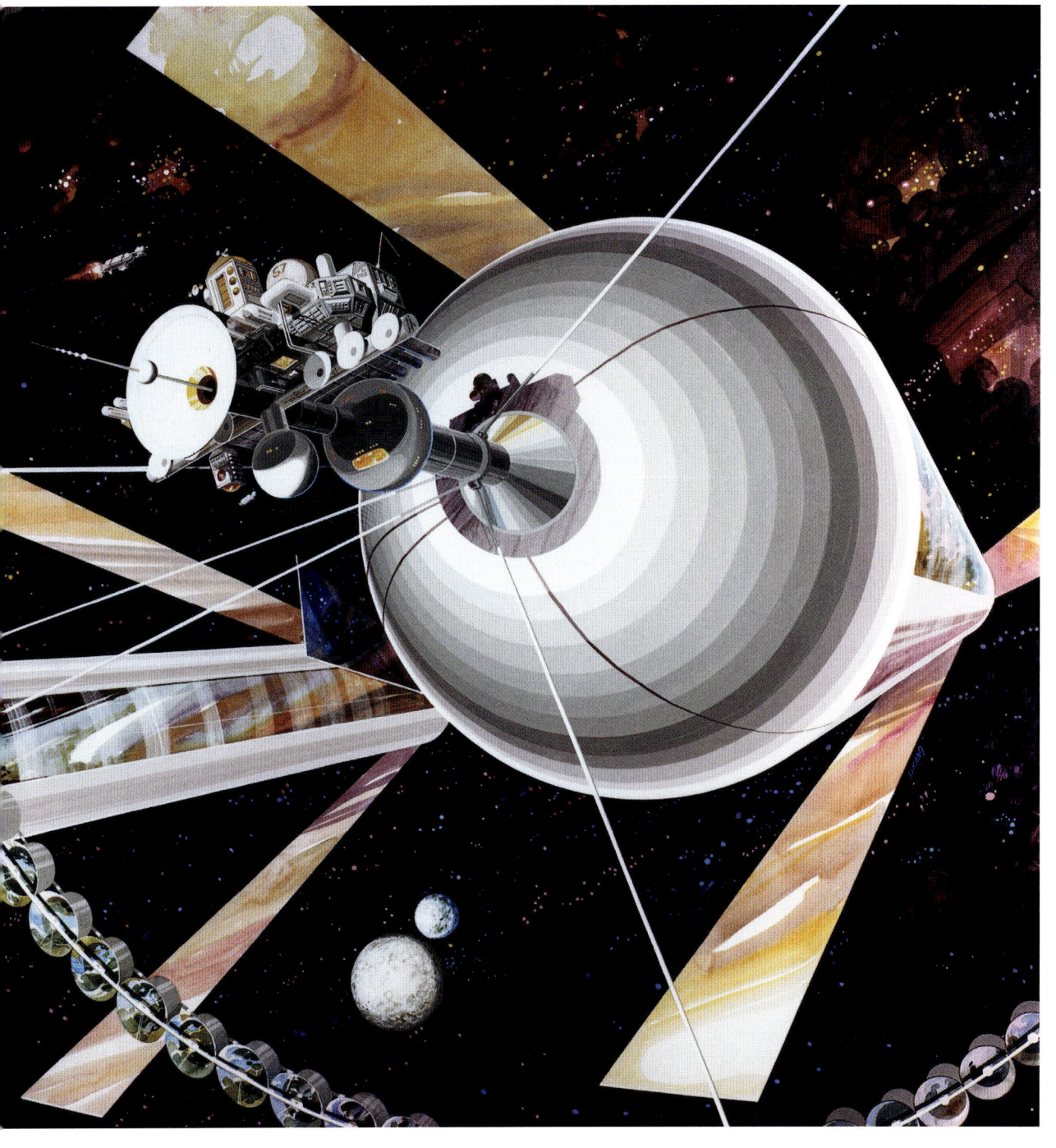

Below: The O'Neill cylinder, a space settlement concept proposed by American physicist Gerard K. O'Neill in his 1976 book, *The High Frontier*.

Below: Space Frontier conference in the 1990s, attended by Elon Musk, who at the time was looking to do something in space.

Right: NASA's Space Launch System (SLS) rocket with the Orion spacecraft atop launches from Launch Complex 39B at NASA's Kennedy Space Center in Florida, November 2022.

WALT ANDERSON: I had a company in Washington, DC, called Mid Atlantic Telecom, which provided long-distance services to businesses. The company was started in 1983, and although it was never a billion-dollar enterprise, it was growing pretty rapidly. In 1988, the company had about 35 employees and was doing OK. That's when I learned about the International Space University [ISU] and went to meet the guys who were trying to start it up. I got excited, and they invited me to be part of the team.

A few years later, I sold my company for a total of around $22 million, of which I received half. I started another company in Europe and was involved in another company in the USA. They both became unicorns within a few years.[2]

Rick [Tumlinson] was introduced to me around the same time the ISU was getting started. A long-time space advocate, Rick and two other people were founding the Space Frontier Foundation. I met him in the bowels of this huge old aircraft carrier in the port of New York[3] in a little room that he had rented for his office, and he told me about the Foundation. The Foundation was his vision of a not-for-profit organisation that would bring together people who wanted to do commercial human space flight. Rick was radical by definition, being somebody who was willing to swim upstream to promote the things he believed in.

---

[2] A unicorn is a company that goes public at a valuation of more than $1 billion.

[3] The USS Intrepid, a WWII aircraft carrier is moored today at Pier 86 on the Hudson River in New York, as a museum of Air and Space.

RICK TUMLINSON: At the Space Frontier Foundation, we were the bad boys and girls. We were the radicals. We believed that the government's role in space is funding long-lead technology development, pure science, and what you might call the Lewis and Clark function of long-lead exploration. Their job is to go over the hill and tell us what's there, map it, and then get out of the way and let the people take over. But what was happening was they weren't letting go. They wanted to own everything, top to bottom. It was a model they had created for Apollo, and it had worked for that. But it wasn't going to work to open the frontier, so we basically went to war with the establishment. And Walt was attracted by what we were doing, because he is a rebel by nature.

Gardellini attended the ISU summer programme at MIT, as a recent graduate in his early 20s.

CARLOS GARDELLINI: We had just had finals at MIT and were burning off steam, so I wasn't prepared to meet someone who would be such a significant player in the events of my life. Someone said, 'One of the biggest benefactors of this programme is here for the Fourth of July party, do you want to meet him?' I pictured someone in a three-piece suit, clean cut, very conservative. Walt was working the bar, doing what he does best: [he can] size you up in five seconds and tell whether you're a useful tool or an arsehole or, God forbid, a weasel. And for whatever reason, in my five seconds under his laser focus, he saw in me a tool. By September, I was working for one of his space companies.

All the people in 'commercial space' with all these great ideas to build reusable rockets, and new kinds of satellites that could be refuelled in space so you can save money, and ground support equipment, and antenna systems, and on and on and on, we collectively called 'New Space'. All of us in New Space firmly believed that NASA had let us down, and the only way we were going to get [to space] was by us doing it ourselves. That's dogma 101, baby. 'Old Space' is these huge companies that have been government contractors for decades. That was the only game in town if you wanted to work in space. And I worked that side of the street for many years.

RICK TUMLINSON: The concept of New Space was to call out, to name, to identify, to evoke this idea of an entrepreneurial, creative, driven type of entity that was very adaptable and would be able to begin replacing the aerospace–industrial complex and open the frontier. So we created the Foundation for the International Non-governmental Development of Space as a way for Walt to manage his donations. Walt brought in Carlos as a third vote. So it was basically the three of us who would make decisions on where the money went. And Carlos was great to work with. He has an intuitive, financial, pragmatic side to him when he's examining these kinds of things.

CARLOS GARDELLINI: New Space is akin to Silicon Valley. Crazy entrepreneurs with insane ideas, and they band together and change the world. The difference is, in space it's very difficult to succeed, so that makes you tighter with your peer group. It was a special time in the sense that instead of working for one of the major military–industrial complex corporations, you had the chance to work and play and associate with fellow enthusiasts who had tossed aside guaranteed employment. Walt, God bless him, tried to bend over backwards to accommodate many a crazy-ass idea.

Left: Carlos Gardellini and Walt Anderson sightseeing in the 1990s.

Below: Richard Garriott at his desk in New Hampshire in 1984, creating Ultima IV, his first bestselling computer game.

WALT ANDERSON: I was pretty well known [in the New Space community], because I was a source of funds, and funds were desperately needed. Whenever I attended conferences, people would come to me with notebooks full of pictures of all the things they were never going to build that they wanted to get funded, some of them practical and some not. There were ideas whose time had not yet come; for example, a space elevator, the idea being that you put a cable between the Earth and a low-Earth orbit. It's a great idea – Arthur C. Clarke wrote about it in The Fountains of Paradise – but a space elevator is only needed when you have enough stuff to elevate. Nobody was going to put $1 trillion into a space elevator 20 years ago. So, the Space Frontier Foundation attracted everyone, from those who were a little bit nutty to people who wanted to build practical stuff and make a difference.

At these conferences, you met people with different goals and people with genuine technical expertise, so they were useful places to find partners, collaborators and ideas. After he sold PayPal, Elon Musk came to a couple of Space Frontier Foundation events. He had decided he wanted to do something in space, and he was looking for the path that he wanted to take. I don't know if he got his inspiration from Space Frontier, but he came a couple of years in a row.

RICHARD GARRIOTT: When suddenly we got stuck in low-Earth orbit with the space shuttle, which was expensive, slow and dangerous, two things happened. One, everybody was profoundly disappointed, but we were also inspired to go do other high-tech stuff. Whether it was someone like Elon Musk with cars or Jeff Bezos with Amazon or me with video games, I would argue that the tech boom was a direct ancestor of the generation inspired by Apollo. We all went off to become entrepreneurs and earn some money, then we all came back and said, 'Let's go fix that space thing.'

THE NEW SPACERS   191

STRUGGLES IN THE RUSSIAN SPACE PROGRAMME

The fall of the Soviet Union placed significant financial restraints on the Russian space programme that had a material impact on cosmonaut-in-training Sergei Zalyotin.

SERGEI ZALYOTIN: I enrolled in [the cosmonaut programme in] 1990, and exactly one year later, in December 1991, the Soviet Union unfortunately crumbled. As a young cosmonaut, I saw what was unfolding but from a distance, from behind the fence. This was because I had a pretty tight schedule – four double lessons every day. I watched events on TV and heard about it all, but I didn't take it seriously. I only realised what was happening six months later when intense inflation set in. And when my cosmonaut salary fell, then, of course, times were quite difficult.

To ensure my family was financially secure, I had to work part-time as a taxi driver for a few months. I'm not ashamed of that, because I didn't want my children and wife to suffer. Any decent person would fight to provide for their family. For a time, many Russian cosmonauts worked second jobs. Almost my entire unit did this.

If there are problems in the country, of course this is going to hit particular industries more. Underfunding had a huge impact on our space exploration programme, so for the next 15 to 20 years we slowed right down and stayed at the same level of development as during the final years of the Soviet Union. A huge number of scientists went to other industries where they could make better money, and many moved abroad.

Below: Russian Mission Control Room near Moscow during the 1990s.

American Vice President Al Gore and Russian Prime Minister Viktor Chernomyrdin announced plans for the International Space Station in September 1993, which would be a joint venture between five space programmes and fifteen countries. The first module was launched in November 1998.

SERGEI ZALYOTIN: The Americans had already begun to squeeze the Russian space programme, and [our] flights to Mir were being postponed because of underfunding. We had debts for the International Space Station, and they [NASA] were saying to us, 'How can you have your own station when you owe us money for the ISS?' They wrote off some of the debt, but this was on the condition that we got rid of Mir. What's more, they made it into a kind of PR campaign. They said that there should only be one big international station, and no other stations nearby.

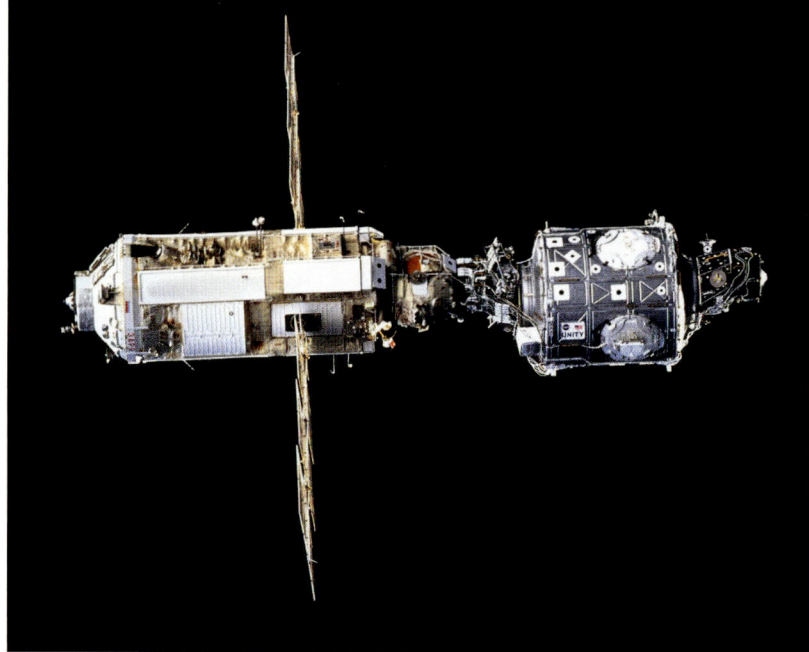

Top left: Al Gore and Viktor Chernomyrdin announce plans for the ISS at a press conference in 1993.

Bottom left: The first two pieces (Russian-Zarya and US-Unity) of the ISS are mated together, of what would become the largest manmade item in space.

Right: Astronauts Jerry Ross (left) and James Newman work together on the final of three space walks of the STS-88 mission. Featured is one of the solar panels of the Russian-built Zarya module.

Overleaf: Space Shuttle Endeavour arrives at Launch Pad 39A in the dim early morning light, atop the mobile launcher platform and crawler transporter.

MIRCORP

Right: Mir, as seen from the US Space Shuttle in 1998.

RICK TUMLINSON: In the late 1990s, Walt was investing in a lot of start-ups, and my job was to give away roughly half of the value of the increase in those holdings. At that time, he had invested in maybe a half a dozen companies, and I think three or four were going to go public, so we were confident that we were going to be sitting on a very sizeable amount of funding very quickly. Instead, several of those holdings got wiped out in what was called the dotcom meltdown.

The Space Frontier Foundation was pushing for the creation of whatever we could get in orbit to begin opening the frontier when we heard about the plan to bring Mir down under Yeltsin. I think it was at that point one of our members, a guy named David Anderman, came to me and said, 'Why don't we try and convince him not to deorbit it?' So, we started a project called 'Keep Mir Alive'. And I actually wrote a letter, which I still have, to President Yeltsin, asking him in the name of the pride of the Russian people to keep the station going.

I then had a meeting in Houston and then got talking to Pete Worden,[4] who had just got back from Russia. He mentioned a price for buying a ride on Soyuz. I'd never heard numbers so low before. I started doing the math, and I was thinking, *I've got this rich guy named Walt Anderson over here, and all of these investments are about to pop and we're going to have lots of money, and David Anderman has said we should figure out how to save Mir. Why don't we buy it?* So, I reached out to Walt and said, 'How would you like to own a space station?' He said, 'Great.' We then had lunch in an LA restaurant with those paper tablecloths, and we started drawing up ideas of what we could do with Mir.

WALT ANDERSON: Rick Tumlinson apparently had some conversations with people in Russia who said, 'If you want to save the Mir, you've got to pay for it.' And he came to me and said, 'What do you think about the idea of saving Mir?' And I said, 'This is crazy. Tell me more.' I mean, who doesn't want to own a space station?

We never actually owned it, but we did what's called a triple net lease, which means that if you rent a building, you are responsible for providing heat, cooling, decoration, everything. You get the building and you're totally responsible for it.

Our goal was to do not only scientific research on Mir but also industrial research; for example, if somebody wanted to look at processing things in zero or low gravity. We also saw it as a commercial opportunity. The film director James Cameron had a type of 3D camera system that he wanted to bring up to Mir. We said, 'If you pay us, you can do it.' And the producer Mark Burnett wanted to do a series of *Survivor* on the station. So, we had a lot of ideas about what we could do with Mir and make it a commercial asset.

---

[4] An astrophysicist who held a number of positions in the US Air Force before later becoming the director of NASA's Ames Research Center.

CARLOS GARDELLINI: I was living a semi-retired life in a water-facing condo in South Beach, Miami. I was on my balcony and the phone rang. This would have been late summer of 1999. It was Walt, and he literally opened the conversation by saying, 'I'm thinking about buying a space station. I'd like you to help me.' Just like that. No introduction, no preamble, no anything. Who in their wildest dreams imagines that when someone says that to you, they're talking about an actual space station that's in orbit over Earth? I thought, *He's got some crazy idea that he wants one of these things transported somewhere, like a museum.* That's when he drops the bomb and says that he's talking about the Mir space station that has actually been flying up there for nearly 15 years.

I've known him at that point for just over a decade, but you don't get a phone call every day saying, 'Hey, let's buy a space station.' How do you answer that? Well, this is one of those times you've got to say, 'Yeah, sure, how can I help?' We didn't have a name for it, and we had no idea it was going to be a business, but we more or less pencilled together all of MirCorp [on that call].

In the back of my mind, during that first phone call, I was thinking of saying, 'Clearly, the first thing you do when you buy a piece of real estate is conduct a site survey. I'm your man. Send me up there.' But I didn't. I've always regretted that. I don't know what his response would have been, but, you know, stranger things.

Somebody had to go out and meet the Russians and have preliminary discussions. Can they work with a Western partner? Are they willing to meet Walt's terms? Which were no corruption, no bribes, no oligarchs, all on the up and up, all transparent. The idea is we would flip this on the commercial markets as quickly as humanly possible, and if there was even a whiff of corruption, they wouldn't allow us to go public.

Jeffrey Manber, who spoke fluent Russian, was an American living in Moscow working for Energia, which was, essentially, the Russian space programme. Think of them as NASA, but if the United States stopped existing and the only way NASA could continue is if they listed themselves on the New York Stock Exchange. They did Sputnik, they did Yuri Gagarin, they did the first woman in space, as an entity of government, but now they were a company. The first trip, everything came down to Jeffrey, my compatriot Rick Tumlinson and me.

In the Soviet system, you start at the ground level and meet with the third-level engineers, and then if you pass that test, you meet with the next level, and then if you pass that test, you finally meet the big guy, Dr Yuri Semenov, the head of Energia and therefore the de facto head of the Russian space programme. He was the decision maker who was going to determine if this had legs or not. And we got to Semenov, I don't know, third or fourth day. Why was that? Because they'd already done a deep look at Walt Anderson and figured he was worth somewhere between half a billion and a billion.

We met with Semenov [in] the first week of October. We were ready to sign when Semenov grabbed both contracts and crossed out the dates and wrote 30 October and said, 'If it's such a great idea to meet again on 15 November, then it's a better idea to meet two weeks earlier.' We walked out of that room, and I thought, *What are we going to say to Walt? He's going to kill us. He's probably meeting with some prime minister on that date.* To his credit, Walt said, 'That's fine. I'm just going to cross out the seven things I was doing that week.'

Above: Mircorp private jet trip to Moscow in the 1990s.

Below: Carlos and Rick pose with Mir model in Russian Mission Control Centre in 1990s

WALT ANDERSON: I went to Russia after Rick and Carlos had arranged a meeting. We took a private G-IV jet, because we wanted to impress and show that we were real and had some resources.

CARLOS GARDELLINI: We were on this G-IV, which is the ultimate projection of success and power. You just show up at the tarmac and they greet you right there at the door to the plane – no hassle, no bustle. Only way to fly. We were taking all these photos, because we were excited, then a dude showed up from Domino's and delivered four pizzas. That's the essence of Walt Anderson: it's not about the money or a projection of wealth. It's not about being; it's about doing.

Halfway across the Atlantic, Walt brought out the board game Risk. So, we were on a G-IV, eating pizza, drinking private-label Walt Anderson champagne, playing a game of world domination. It was beyond anything you could believe.

Pretty much everybody on that plane thought it was a foregone conclusion that the world would be a spacefaring civilisation eventually, and this was the opportunity of all of our lifetimes. Again, it was jumping the line, getting there much faster than any other way. It was beyond anything we could have imagined to this point.

[In Moscow,] we went through all the preliminary stuff: the Russians giving Walt the tour of all the assets and all of the incredible things the Russian space programme was doing and meeting with all the key people. Then, towards the end of the three or four days, it came down to the big meeting between Semenov and Anderson to see if we could reach an agreement to save this thing. We had to get them on the same page. And, amazingly enough, the meeting of minds occurred.

202 THE NEW SPACERS

Below: Director General of Russian space agency Yuri Koptev shakes hands with the Director of RSC-Energia Plant Yuri Semenov during a signing ceremony in April 1999 for the construction of a module for the ISS.

WALT ANDERSON: Semenov was initially brought in as an apparatchik under the Soviet system. The government would send people in to monitor large companies to make sure they were following the party line, and that's how he started. But he was a smart guy, and he became a space enthusiast and somebody who was willing to challenge the government and try to protect his organisation. So, he was a real advocate for saving Energia and the development of space for peaceful purposes, non-military purposes. He was very much aligned, oddly enough, with my ideas. That's why he was willing to let a bunch of crazy Westerners fly in and talk to him. He would have done anything to save Mir.

The geopolitical relationship with Russia and America has always been fraught, but we weren't there as Americans. We were there as developers of space for all mankind. That's the approach we took, and the Russians responded in kind, so we didn't have to deal with geopolitics. We dealt with the practical possibility of operating a business to save Mir.

I had prepared to send a deposit of $7 million to show good faith. It was right before the holiday season, and we had information that Energia might not meet their payroll needs at the end of December. So, it was a good time to send them a chunk of money to keep them moving forward.

I had talked to my banker in advance and arranged that if I called up and gave him a code word, he would send the funds. By saying that code word, the $7 million went to Energia. I'm not sure the Russians really believed that money was heading their way until the next day when they got it. It was a dramatic gesture, but once I'm ready to do something, I want to move forward quickly, so I was eager that they would get that money and begin the planning process in earnest to do what we needed to do, because it was a huge task ahead.

CARLOS GARDELLINI: It's of no value if you have the world's greatest hotel, but there are no roads to it. MirCorp wasn't just the destination; MirCorp was the road. Soyuz capsules would take you there and safely bring you back. This was the whole package. The deal with Energia was that they would handle all the hardware, including the system to get there and back.

RICK TUMLINSON: When the agreement was signed, it was surreal. There was a moment of triumph, feeling like we were changing history, but terror at the same time. I remember thinking, *I've got to get to work on this*. When you have this grand idea and it's going to change everything, that's great, but then you start getting into the reality of it.

CARLOS GARDELLINI: We went to a fabulous restaurant in Moscow, and all the key players were there. Walt said, 'Write down what percentage chance you have for this project succeeding.' Pretty much everyone wrote down a number that wasn't huge, but the lowest number until I went might have been about 30. I wrote '0'. But at that point, we had made our bed, and we were going to sleep in it.

Left: A MirCorp meeting with Moscow officials.

Below: View through the Zarya module window on the ISS, taken during Expedition 13 on 5 August 2006. A Soyuz spacecraft can be seen docked to the station.

THE NEW SPACERS 205

THE FIRST PRIVATE SPACE MISSION

CARLOS GARDELLINI: The first and, let's face it, only [MirCorp flight] was to evaluate the status of Mir as an ongoing enterprise and what it would take to ensure that we could keep it going. This was the first fully funded private orbital mission. This was the real deal.

[We felt that] it was absolutely in trouble, but salvageable. [It was made up of] six or seven key modules, of which two or three were relatively new. They weren't a problem. But [we knew] some of the core modules were approaching 15 years old, and the one that was hit by Progress was leaking. So, first thing you want to do is plug the leak. Then you move forward: what's the next thing you do? And the next thing? You have to follow the checklist.

Right: MirCorp's first launch on a Soyuz rocket carrying cosmonauts Sergei Zalyotin and Alexander Kaleri.

Zalyotin made the first of his two flights in space on Soyuz TM-30 on 4 April 2000, the mission paid for by MirCorp.

SERGEI ZALYOTIN: As far as I was informed, the situation with Mir Corporation [MirCorp] was that they offered to fund the station. And from the point of view of ideology, not from the point of view of technology, it would come under their control. In other words, they would fix it up, operate it and pay for it. They would become the owners of the station.

I had mixed feelings about Mir being claimed by the Americans. On the one hand, it was disappointing that we couldn't [fund it ourselves], because Russia was such a huge and powerful country. On the other hand, I had the chance to fulfil my ambition of flying into space. Because the worst thing for any cosmonaut or astronaut is to train for 5, 10, 15, 20 years and not fly into space. That would be a great tragedy.

I understand that for this company to buy and use something that belonged to Russia and the world as a whole, was a huge PR bump for them. But taking people up there was an impossible task. What would they be flying these people on? On shuttles that were no longer flying to Mir?[5] On Russian ships? There would need to be at least one experienced cosmonaut to take two tourists there at a time, but one of the tourists would need to be trained for at least two or three years so that they could understand which buttons to press. From a technical point of view, space is only for people who are trained. Not every ordinary person on the street can go on a spaceship. You have to be prepared.

We flew out purely to revive the station, because there was a fear that the pressure would drop, and then no one could fly there anymore. The two-and-half-month mission was a historic one, because I knew that it might be the last ever trip to Mir, and I really wanted to fly to my home station. But It was a very challenging mission, because the station was nearly 15 years old.

After we docked and checked everything was air-tight, we couldn't open the hatch. The station had not been operational for about nine or ten months, and the rubber that acts as an airtight seal was stuck together. No matter how hard we tried, we couldn't open it. Thank goodness I had an experienced flight engineer who managed to back up and kick the hatch in.

[From our inspection] I concluded that the station was fully maintainable and that we could operate it for at least three to four more years. It was full of scientific equipment, and we were ready to do it. But then politics intervened.

---

[5] The final docking mission of the Shuttle-Mir program took place in June 1998. This was the last time a Space Shuttle visited the Mir space station.

Left: Zalyotin's personal footage from the MirCorp mission to Mir in April 2000.

Above: Kaleri's footage showing Zalyotin in front of the Russian flag, April 2000.

Above: Zalyotin exercising on Mir, April 2000.

Above: Floating Russian dolls of Zalyotin and Kaleri (the mission's flight engineer), April 2000.

CARLOS GARDELLINI: That first mission succeeded, and it was determined that, political risk aside, this [was] a going enterprise. The problem [was], you can't say political risk aside, because, as I saw it, this project was 100 per cent political risk. The powers that be were going to make sure this thing was thrown in the ocean. Full stop.

WALT ANDERSON: Right before the launch, the Russians called Jeffrey Manber, [who was now] the president of MirCorp, into the control room and said, 'We no longer control Mir. You have to give permission for your private launch.' So Jeff was the one who ordered the launch of the private rocket that we had leased to the private station to do the survey for the renovation.

I provided the funds to MirCorp – [the mission] turned out to be $20 million – so it was a private mission, even though it was operated by Energia from their control centre. We were effectively buying the rockets from Energia and when they launched, they were our rockets, and although they were trained Russian cosmonauts, they were employed by us.

We did our first commercial activity on that mission. RadioShack had come out with a talking picture frame, so we had the family of one of the cosmonauts record something on the frame, and it was handed to him while it was being filmed. RadioShack paid us for that commercial to be filmed on the ISS.

RICK TUMLINSON: I was in LA giving a briefing at the Space Tourism Society lunch, and I told them about MirCorp. John Spencer, who founded that organisation, came up to me and said, 'I want to introduce you to Dennis Tito, who has been wanting to go to space.' Dennis and I had a conversation, and I met him twice for lunch. I said, 'Do you really want to go?' He said, 'Yes.' I said, 'I think we can make that happen.' He then wired a $1 million deposit to MirCorp and the process began.

WALT ANDERSON: Soyuz capsules had three seats, and the idea was you'd have two cosmonauts and one tourist in the third seat to help pay for the missions. We had Dennis Tito signed up to be the world's first space tourist. Tito was a fund manager, and we negotiated a deal for him to pay $12 million to fly to Mir.

Below: The Space Shuttle docking target on Mir.

THE BACKLASH

WALT ANDERSON: [A journalist] came to us and said that she wanted to do an article about Mir – not about me – and that made sense. It was a great story. So, we allowed her to travel with us and brought her behind the scenes. If you were a person who wrote about space, you would have killed to go on that trip, but she did this report that made fun of everything we were doing: 'Wacky Walter, anti-capitalist of space'.

CARLOS GARDELLINI: It was just a reporter following a story, and it led her to this conclusion. There were certain people in the space community who might not have liked her tone, and who felt that she hoodwinked Walt into thinking that she was part of the team, but she was just a reporter following a story.

RICK TUMLINSON: What ended the project was less about the public reaction and more to do with political decisions on the part of the Russian government to stay in the good graces of NASA, which was paying a lot more money than we were. Walt had some financial problems right at that point with the dotcom meltdown, but we could have worked our way through those financial challenges if the pressure wasn't being put on the Russian government.

MICHAEL FOALE: In the late 1990s, Russia had made a commitment to deorbit Mir in 2001. NASA, at the top level, wanted this to happen. They wanted the Russians to stop putting resources into sending rockets to Mir and focus on getting the Russian pieces of the International Space Station ready. It was all to do with American perceptions of how much money the Russians would put into Mir, as opposed to into the ISS. They wanted Mir to go away.

WALT ANDERSON: NASA saw MirCorp as competing for Russian resources that they were using to build the International Space Station. It would have been a bit embarrassing if our $100 million investment in MirCorp renovated Mir to become more modern than the $100 billion, overbudget ISS. But we were never anti-ISS. We said we wanted to have our own resource to do things, because the ISS at that time was very closed to commercial activities. So, we were opening up the opportunity for people anywhere in the world to use the renovated Mir as a commercial platform to do things in space.

CARLOS GARDELLINI: The Soviet Union collapsed and left 15 nation states even worse off. At least when they were together, they had something. Now, all of a sudden, you have 15 independent, completely bankrupt countries looking for anything to survive. Kazakhstan has got one hell of a potentially dangerous arsenal of missiles. You'd kind of like to control that, wouldn't you? And one of the ways you control that is by keeping the engineers gainfully employed on peaceful projects like the International Space Station. So, I don't have any misgivings that the United States had an approach that probably made sense. For very small amounts of money, toss Mir into the ocean and build this unbelievably expensive replacement. Sold. And then along come these anarcho-capitalists with a different idea.

Above: A view from Mir, taken during the MirCorp survey mission.

RICK TUMLINSON: NASA freaked out when we did this. They hated it. They took it as a direct threat to what they were doing. They were trying to characterise us as somehow being un-American for what we were doing and [to] portray Walt as being an anarchist. I equated it to being thrown in a pool full of sharks and trying to inflate a lifeboat while somebody's shooting at you with a machine gun in terms of trying to pull this whole thing together. The Russian establishment was coming at us, Congress was coming at us, the media was coming at us. It was crazy.

The crux of NASA's argument against what we were doing wasn't really clear. It was more of a flailing reaction of this giant beast that was being challenged. There were different arguments on different days. They were throwing out whatever it was they could do to discredit us. It really boiled down to, 'How dare you?'

SERGEI ZALYOTIN: The guys [cosmonauts] who were going to America a lot during that time saw that activities on the Russian Mir station were not shown or discussed in the US press. There was lots of talk about the ISS, but there was nothing being said about Mir in the press or on television. For the Americans, Mir was an eyesore. They simply didn't need it.

Left: Sergei Zalyotin on the MirCorp mission.

Below: Sasha Lazutkin and Vasily Tsibliyev on Mir in the 1990s.

SASHA LAZUTKIN: I had a clear understanding that Mir was being deorbited only because the Americans wanted that to happen. That was unequivocally the case. Mir was a competitor to the ISS.

The process was underway for the construction of the ISS, and, naturally, there was a discussion as to who would take the lead on that station. Our people said, 'There is the mission control in Moscow, and there's the one in Houston. Why not have two?' The Americans said, 'No, there should be only one, and it will be Houston.' What language would be the principal one used for communication? Our people said, 'When we flew missions with you before, we spoke in English and Russian.' But the Americans said, 'No, the language will be English.' We wanted to build the International Space Station on equal terms, but they said, 'If we're on equal terms, a decision will need to be taken, and we will have to coordinate that with you. It will be better if we take the decision as the lead and you implement that decision.'

If Mir had still been orbiting, a situation could have unfolded where we might have argued with the Americans and said, 'Look, you be in charge of your station. We'll walk away and conduct our own national space programme.' You see, Mir performed the same functions that were planned for the International Space Station. Therefore, to avoid that situation, the Americans needed us to deorbit Mir.

THE NEW SPACERS 217

THE DEORBITING OF MIR

WALT ANDERSON: I think we found out about Mir being deorbited shortly before it was publicly announced, too late for us to attempt any intervention where we might have flown over and tried to convince them to not do it. We were quite upset when we realised we didn't have time to take any of the normal steps where you go lobby and intervene and meet people. It was already a done deal.

MirCorp obviously still existed, but we had lost all the money that I had invested. We did continue operation to meet our obligation to Mr Tito to make sure that he got a flight on the ISS, and I later attended his launch. After that, I felt like MirCorp was done. We had the opportunity to continue to be a space tourist agency and sell Russian flights to the ISS, but I felt that the bigger opportunity to make a real difference in terms of a private space programme was gone.

RICK TUMLINSON: I was in California when [Mir] was deorbited, and it was a very sad day. We really thought we were on to something, and it also brought home to me the scale of the forces that we were up against at that point. But you never give up. You've got to keep going.

SERGEI ZALYOTIN: In March 2001, when Sasha Kaleri and I were offered a trip to Fiji to watch and comment on the deorbiting of Mir into the ocean, we refused to do so. For me, it would have been like watching my home burn down. Mir was our Russian two-room apartment, which I really enjoyed living in. I was the last commander of the Mir station. I loved it very much, so I didn't want to watch it come down from orbit and sink. I had tears in my eyes the first time I saw the footage of the sinking of Mir.

It was disappointing that our country was unable to preserve and extend the lifetime of our pride, our beauty, some cosmonauts even call it our Mother in Space, the Mir station. The International Space Station was almost finished, so we were already paying more attention to that. Yet any of us who watched the house where we were born as cosmonauts being destroyed felt very sad about it, no doubt about it.

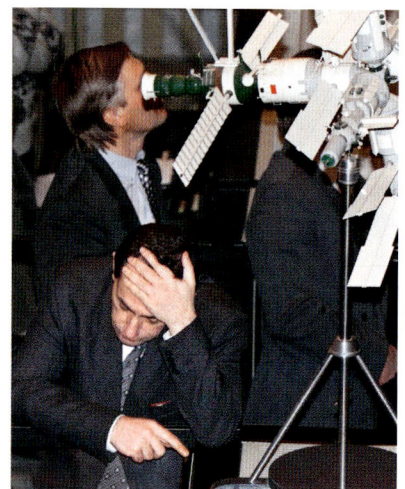

Above: A Russian space mission controller looks away as his colleagues watch the controlled re-entry of Mir into the atmosphere, crashing down into the Pacific Ocean.

Right: Mir re-entering the atmosphere, over the Pacific Ocean on 23 March 2001.

SASHA LAZUTKIN: When they showed Mir being deorbited on TV, there were mostly Russian government officials commenting on this event. They said the station was old and constantly breaking down, and compared it to a car. They said that if you'd had your car for 15 years, that meant it was old. But the Mir space station was not a car. The station was a home. You don't demolish your home when the fridge or the television plays up. You take that equipment away and fit new models. The walls remain. Things will play up on the other space station too. That doesn't mean that the station is old.

The Mir space station was the symbol of our national space programme, so when it was deorbited, there was no applause at Mission Control. We never discussed what the station meant to us, but when it was downed, each department held a wake. Not a celebration of a successful end to the project, but a wake for our own national space programme. We were told there would be new opportunities for us, that there would now be an international programme. But that entire international programme would be tied to one thing: to take our experience. Which experiments to perform, what results could be obtained at the station, and so on and so forth. How to build the new station.

Only we knew how to do that. How to operate a large station. Only we knew how to do that. How to conduct long-term missions in space. Only we knew how to do that. And we handed all that experience over to the Americans and Europeans. We gave away our knowledge, while we ourselves stayed where we were. Yes, there was international cooperation, and we were working on how to collaborate in space. But the main issue was that our national programme had been shut down, and we did not move forward.

Left: US space tourist Dennis Tito (left) and Russian cosmonaut Talgat Musabayev (right) look out from the Soyuz capsule after their landing near the Kazakh town of Arkalyk.

Right: The first segments of the ISS, orbiting the Earth in free fall at 5 miles a second.

MIRCORP'S LEGACY

CARLOS GARDELLINI: I've had two and a half decades to process all of this, and I think it played out the way it had to play out. It was a great, audacious dream, but it wasn't meant to be from the very beginning. We could question whether it should have even been attempted, but we're now where MirCorp was 25 years ago. All the dreams that we had, these guys have followed up on that dream, and it's working. Jeff Bezos is sending people to space. Elon Musk is sending people to space. The dam has burst; there's no going back. Maybe you need to have the pioneers who take the arrows in the back, and it's the people who come after them who establish the hamlets and the societies and the cities.

Initially, when the thing went in the ocean, I thought, *We did all this for nothing.* Since then, [I've become] more upbeat and optimistic about space opening up than I was on 20 July 1969. I say that hand over heart. NASA gets it. They are supportive of commercial companies. I'm not saying it will be easy, but this is what we were hoping for at the turn of the millennium. We don't want to shut down NASA. In fact, if anything, we only want to give them more money. But we always wanted them to turn over a small portion to private industry. Let private industry do this part, and let NASA travel to the Moon, Mars and the cosmos at large.

WALT ANDERSON: Look at all the things ISS is doing now that they weren't doing in the beginning, such as allowing commercially related activities. All those things could have been done on Mir, probably a little more efficiently, and at a little lower cost because of our more commercial view, but they're getting done now.

I believe, based on what people have told me, that the backlash from what was done to MirCorp helped make it a little easier for the next company that came along that wanted to do commercial space and not run afoul of NASA. There are now a lot of companies doing that, and some of them are actually getting support from NASA, so their culture has really changed. I hope that there's some positive legacy.

I certainly lost that battle but, ultimately, the overarching goal I had of commercialising space and providing a better opportunity for more humans to potentially be living off the Earth in the future – I think we're closer to that than we were before. And I now feel that it is possible that there will be some real space stations housing some number of people off the Earth in my lifetime.

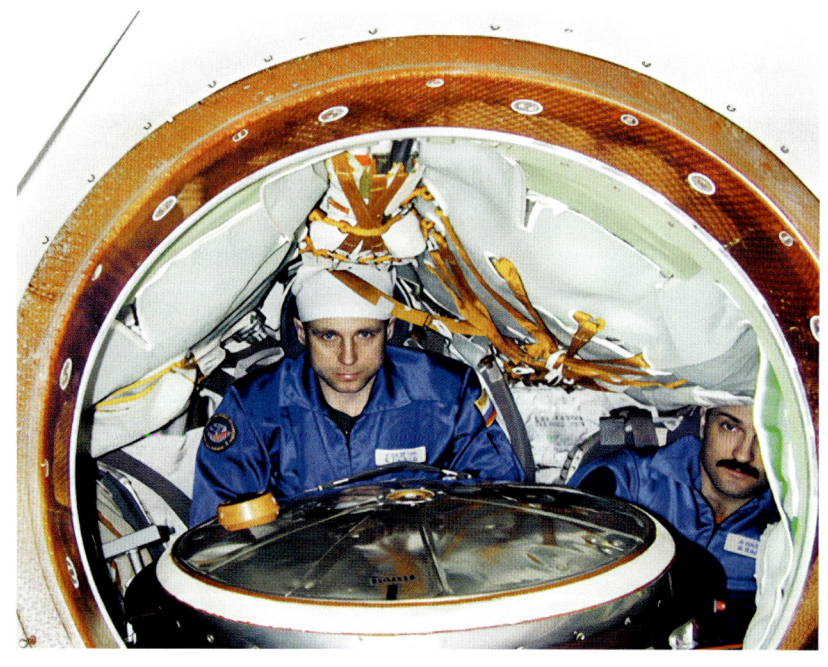

RICK TUMLINSON: I'm currently dealing with several companies who are building commercial space stations, and they all claim that they're going to be the world's first. No, they're actually going to be the second or third or fourth or fifth. We sent a commercial crew to a commercial space facility. That's the facts. My job is to get all of these private companies going, but MirCorp was the first private space facility. And then we got our asses kicked.

I think it was a shot fired in the revolution, and I think it helped prove to some degree that you could begin to take those steps. It also helped to show that there were people out there with money who would back such a venture, and people out there with money who would fly in space.

You can't have a revolution without casualties. Look at where we are now – it could be argued that the private sector is actually leading the way. And I think we're going to get there, no matter what you might think of the personalities involved. I'm always going to be full of gratitude that Walt stepped up. He really put it on the line.

You can never change the past, but sometimes I do think that if we had

succeeded, we'd be in a different place today. I think that it would have been that foundational anchor activity that would have proved we could do things. We'd probably be mining asteroids by now. There might even be people on Mars. If we had been able to execute our business plan, it would have enabled a whole series of activities.

I don't know if MirCorp was too early, but the wall that we were trying to break down was, and still is to a large degree, behind the edifice of our national space programme in the USA. And I hate to put it that way, because I know there are amazing people working at NASA and ESA [European Space Agency] and the other space programmes.

I also get irritated when I see these stories that suggest Jeff and Elon showed up one day and did it. That narrative ignores MirCorp and all of the other projects and organisations who came out early [in the 1980s and 1990s] and started this, who stood for the dream and tried to make it happen. When I look back at what we did with MirCorp, I'm really proud.

Left: Mission on Mir.

Above: The MirCorp mission patch and a sticker, reading 'Bringing the World Together in Space'.

THE PRIVATE ASTRONAUTS

One of the main goals of those championing private human space flight development was to enable access to space via non-governmental means. This was particularly the case for people like Garriott who found themselves locked out of mainstream space travel.

RICHARD GARRIOTT: At the age of 13, our family doctor, who worked for NASA, was giving me an eye test and said, 'You're going to need glasses. I hate to break it to you, but you are no longer eligible to be a NASA astronaut.' Prior to that moment, I had just assumed I was going to go to space like everybody else I knew. I went through what I would describe as being like the stages of grief. Finally, after mulling it over, I said, 'Who's that doctor to be the gatekeeper of space? If I can't go by NASA's rules, I'm going to have to make my own space agency.' Of course, at 13, you don't do much about it. But at the age of 15, I started writing video games, and by the age of 19, I was making a lot of money selling them. And the very first thing I did as soon as I had money was invest in opening up private space flight so that I could go.

Everyone in my hometown in Houston knew that I was the kid who was interested in space and had money, so when anybody was leaving NASA to start something entrepreneurial, they showed up at my door, including Buzz Aldrin, who came to my dorm room at the University of Texas and brought with him a video cassette for his pitch for a company called Starcraft Boosters. I didn't invest in that project, but I did invest in a lot of others, so I became one of the early backers of a bunch of commercial attempts to open up space beyond NASA. The one that came the closest from that earliest era was a company called Spacehab, which was a pressurised laboratory that could sit in the space shuttle's payload bay. They also had a design of it where you could basically fit a double-decker bus in there and take 40 civilians up with you at a time to have an experience in space. I thought, *That's my ticket to space.* We built the lab, and NASA said, 'Great module, we'll put it to scientific use, but we are not flying civilians.'

Prior to joining the Explorers Club [a professional society for promoting scientific exploration], where I served as president, all the investing I did was alongside ex-NASA employees or contractors. It was a ten-year lesson to realise these people might be great test pilots or scientists, but that doesn't necessarily mean they're great entrepreneurs, and it doesn't necessarily mean they're great politicians to help open doors. When I joined the Explorers Club, I met people who were used to going places that were hard to get to. That's where I met the people who became my partners in things like the XPRIZE, Zero Gravity Corporation and Space Adventures, which ultimately opened the door for ordinary people to go to space.

[By the late 1990s,] Space Adventures had people signed up to go on suborbital space flights, but we couldn't predict when we'd be able to start taking them into space. At that time, there was still only two ways into space: NASA and Russia. So, we thought, Let's go ask them again. NASA said, 'Thank you very much for the inquiry, but we're never going to take a civilian to space,' which we expected. When we asked Russia, their answer was more interesting. They said, 'No, we won't take people to space, because to do the study to find out if it could be done

Right: Richard Garriott answers questions at a news conference for Space Adventures on 11 June 2008 in New York.

would cost us a lot of money,' which sounded like a qualified yes to us. So, we asked them how much it would cost. They came back and said $300,000, so I wrote them a cheque. Six months later, they came back and said, 'We'll fly your people for $20 million,' which was exactly the price we predicted it was going to be.

Once Space Adventures had negotiated with Russia to be able to fly someone on the Soyuz [to the ISS], I was actually intending to be that first person. But that's also when the internet stock market crashed, and my ability to pay for that first seat disappeared, so we sold it to Dennis Tito. NASA tried to convince him not to go. They literally sent an astronaut to Russia to tell him that they thought it was bad for the USA if he flew with the Russians. He was very upset about the fact that NASA were actively resisting.

I went through a very brief period of feeling resentful towards Dennis. But he had tried to get himself into space just as hard as I had, and he'd come up against the same series of roadblocks. So, if I had to lose out to anyone, Dennis was the right guy.

It was a historic journey. When Dennis flew, we had broken down the barrier that had existed up until that point of only government-employed astronauts or cosmonauts ever going to space. Dennis, just like Jeff Bezos, Elon Musk and me, grew up under Apollo and believed in that dream but was then orphaned by it not happening. Dennis's flight was the day we reversed that trend and said that we were going to manifest the future that we had believed in when we were much younger, which starts with flying ourselves but ends in the journey to finally make humanity multi-planetary.

RICK TUMLINSON: One of the things we're proud of with MirCorp was that we were able to take the initial steps that led to the first private citizen flying into orbit. It was complex. A success like that has a thousand fathers. What I can say is that I signed Dennis up, and he sent a $1 million deposit. But that doesn't mean that other people weren't involved later. They were. Once it became about him and the ISS, he and his team did the work. And Space Adventures get 100 per cent credit for everybody who followed Dennis. But we started the process, and, at the end of the day, the fact that it happened is what's important.

In 2006, Anousheh Ansari was the fourth self-funded person and first self-funded woman to fly in space. She was also the first Muslim woman to fly into space.

ANOUSHEH ANSARI: I was born in Mashhad, the second largest city in Iran. At my grandparents' house, I'd sleep outside on summer nights because we didn't have air conditioning, which would give me the opportunity to stare at the stars. Even now, when I look at the night sky, it's a place of mystery. We know so little about our universe, or the answers to fundamental questions such as, 'Why am I here? Is there a purpose to my life? Do I play a role in the universe?' I think this curiosity and wish to understand the unknown started when I was very young on that balcony.

Life was pretty normal until I was about 12 years old and the revolution started in Iran. It was the first time that I'd witnessed violence. I saw people demonstrating, yelling, shouting, setting cars on fire, shooting guns, and it was scary. The regime changed, and what came after was not what most people I knew expected – an Islamic Republic. We all thought we were going to have a more free, open society. Instead, as a girl, I had to wear a hijab and keep my hair covered, and our freedom started to disappear. You were constantly worried about someone knocking your door down or putting you in jail. This constant living in fear was exhausting. And then, before I could even understand what was happening, the war with Iraq broke out. So, there were bombings, and all sorts of unrest, as well as shortages of everything from bread to fuel.

Space became this place I would go to whenever I wanted to escape the reality that I lived in. I think this is perhaps another reason why I was so enamoured with it, because I could imagine I was somewhere else completely, on another planet where everyone lived in peace and harmony and loved each other. It was a combination of old Persian fables that my grandmother would tell me combined with *Star Trek*. Space was my refuge.

When you start high school in Iran, you have to choose whether you want to study the humanities or maths and science. I always knew that I wanted to study physics, so I chose science. I also wanted to study physics in college, but I knew this would be a problem, because universities had significantly reduced the quotas for women in the scientific and engineering fields after the revolution. All of a sudden, I saw my options for a future that I had imagined for myself completely disappearing.

Fortunately, my aunt, who was a citizen in the USA, had applied for green cards for us, and after 12 years of waiting, the application was finally accepted. This happened right around the time I was getting ready to go to college, so a new door opened for me – to continue my studies in the USA.

Before that happened, I had big aspirations to become an incredible scientist who invented something so amazing that NASA would invite me to become an astronaut and go to space. That was my whole plan, because Iran didn't have a space programme, and I was determined that I was going to find a way to go to space.

I knew very little about how you could become an astronaut. I still wasn't a US citizen, and I had difficulty even speaking the language. I therefore knew my chance of getting into NASA

Left: Anousheh's first visit to the Kennedy Space Centre, aged 22 years old, with former husband Hamid, 1988/89.

Right: Anousheh Ansari at a space exhibition in the 1990s.

was slim to none, and there was no commercial space programme, so I couldn't imagine any other way. I very quickly switched to thinking, *What can I do to get a job and make some money? Then I'll go back to my plans for NASA.* But I never gave up on my dream. It wasn't superficial. It was deeply felt, and I was adamant that I was going to make it happen.

I remember getting ready to go to work one morning, and Dennis Tito was being interviewed on the news. It was the first time I'd heard about the possibility of a commercial passenger going to space. I thought, *OK, if everything else fails, I'm going to do exactly what he's doing. I'm going to buy a ticket to go to space.* The next question was, *How do I get the money?* The speculation was that Tito was paying $20 million, so I became obsessed with having enough money. Even if I had to save every penny that I had, I was going to find a way. Suddenly, I had this glimmer of hope and a clear path to space.

My husband Hamid and I had a company that defied the status quo of the telecom industry. We joked that AT&T, which was the big telecom company, was the Death Star, and we were the rebel forces building software to destroy it. We started as two hard-working engineers and never imagined that our company would grow so big and so fast. It was acquired right before the big telecom bubble in 2000, and we got very lucky. We sold our company for $1.2 billion, and we were able to imagine a different future for ourselves. I thought, *I'm going to go back and pursue my dream of going to space.* The first thing I did was enrol in a programme to get my degree in astronomy while I figured out what my next steps were.

Around the same time, Peter Diamandis, the founder of the XPRIZE, had read one of my interviews, and he decided to approach me about sponsorship of this prize he was creating. I thought, *Space? Yes, please! Let's set up the meeting quickly.* He told us about his idea of a competition to get people from around the world to build a spaceship and go to the edge of space twice to demonstrate their vehicle was viable. If they made it happen, they would win $10 million. After he left, Hamid, my brother-in-law Amir and I talked, and I convinced them it was a great idea and we should do it, so we sponsored the prize. That's how the Ansari XPRIZE was born.

One of the reasons I really liked Peter was that when he came and met with us, he said, 'I can't promise you that you will go to space, but I promise you that I'll introduce you to everyone that I know in the space industry, and I will do my best to make it happen for you.' And he was true to his word. He introduced me to a lot of people, and I started going to all these conferences and meeting people in the space industry.

We were celebrating the first anniversary of the winner of the XPRIZE competition, and a Japanese businessman [called Daisuke Enomoto] who was signed up to go to space with the Russian space agency and Space Adventures was at the party. As we were talking, he said, 'Anousheh, they told me I can choose who will be my back-up during training. So, if you want to, I would love for you to come and train with me.' I said, 'Does that mean I'm in a corner some place? Or am I actually training with real astronauts?' He said, 'You will be in the programme training with the rest of the astronauts, but you will train with a back-up crew.' I said, 'Yes! What do I need to do?' About four months later, I was doing my medical qualification to be accepted into the programme, and after that, I was on a plane to Moscow and Star City.

Left: Anousheh Ansari takes part in a seat fitting session at the Gagarin Cosmonaut Training Centre in Star City outside Moscow, Russia. They are moulding 'couch liners', that are tailored to the shape and size of each cosmonaut to provide support and protection during the intense forces of launch and reentry.

Right: Private space participants Daisuke Enomoto and Anousheh Ansari have a training session in a Soyuz space simulator in Star City on 3 July 2006.

ANOUSHEH ANSARI: On the last day of training, I was very depressed, because I thought my fantasy was coming to an end. The primary crew was supposed to go into quarantine, and I was flying back to Dallas. On my way to the airport, I received a call asking me if I wanted to go to space: '[Daisuke Enomoto has] failed the medical test. You're his back-up, so if you want to go, you can.' I screamed for like a minute. I cannot describe how happy I was. The driver who was taking me to the airport had to hold his ears because I was screaming so loud.

The news reached Iran, and a lot of people were very excited about it, because for the longest time everything in the news about Iran was negative, and here was some positive news about an Iranian. I'm never shy about talking about my Iranian heritage, because I think it's important for people to know who I am. I embody a lot of the positive attributes of our culture, mixed with a lot of freedoms and things I've learned in the USA – the combination has made me who I am. So, even before I was selected for the primary crew, I always wanted to highlight this part of my story: that we should look at the world not through the lens of what governments are doing, but instead via the fact that every person in every country is just a human with the same basic needs and desires to live a fruitful, healthy life. If we can see the world through that lens, it would not be so easy to dismiss a culture or a whole country because of the actions of their leaders.

I therefore saw my journey to space as an opportunity to talk about Iran and Iranian culture, and I designed my patch, even before I knew I was going to fly, with both the Iranian and US

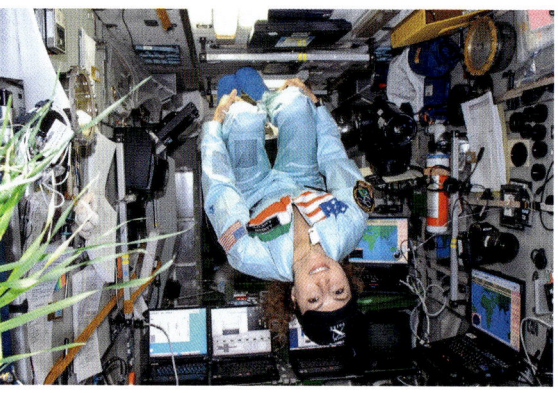

flags on it. When NASA found out that I was going to put the Iranian flag on my suit – and I was only using the Iranian colours, not the Islamic Republic emblem, so it was just a representation of my Iranian background – they reached out through the Russian space agency and told them that I needed to remove it. I was resistant, so during the training I still wore the patch with both flags, and they couldn't do anything. But the night before the launch, someone from the Russian space agency came to me and said, 'We need to remove your patch.' I said, 'This is the spacesuit that has been tested. We know it's functional. You can't just unstitch something – it will create a problem.' They said, 'Well, we can't let you fly with it on.' So, I came up with a clever way of avoiding it: because the Iranian flag is green, white and red, I said, 'Just put some white thread over the green part, and it will look like an extension of the American flag.' I was baffled by the fact that a tiny amount of green thread on my spacesuit was threatening to American sovereignty.

They said it was because there were 15 countries who were members of the ISS, and Iran wasn't one of them, so I couldn't wear the flag unless all 15 countries said yes. I think they were really afraid that I was going to promote the Islamic Republic when I was in space, even though I kept telling them that was the furthest thing from my mind. The only reason I wanted [the Iranian flag on my patch] was to emphasise the whole notion that when we are in space, we're all one nation. That we're all Earthlings who live on one planet.

One of the pictures that many newspapers used was of me before boarding. I had my spacesuit and headset on, but I wasn't wearing the full helmet. The headset covered part of my hair, but not all of it. The picture was used everywhere, even on the Iranian Space Agency website, and people noticed that they had allowed a picture of a woman who didn't have her hair covered on a government website. The space agency quickly realised what had happened, and they edited the picture so that my helmet was covering my hair. It just shows the absurdity of the whole thing. It was the same thing with the little green patch on my spacesuit threatening the American government. Now it was a few strands of my hair threatening the Iranian government. If your ideology is so weak that it's threatened by little things like that, maybe there's a problem with the ideology.

Left: Anousuheh floats in the Russian Zvezda Module onboard the ISS wearing her flightsuit, stitched with the Iranian flag.

Right: The final photo-op for the crew of Soyuz flight TMA-9 to the ISS, before they board their spacecraft. (from top to bottom) Anousheh Ansari, NASA astronaut Michael López-Alegría, and Russian commander Mikhail Tyurin.

THE NEW SPACERS 231

ANOUSHEH ANSARI: I remembered lying on my bed and looking at the stars and imagining myself flying to space. And here I was actually doing it. But it didn't resemble any of my dreams, because they included far more advanced technology; for example, an USS Enterprise-type spaceship. Instead, I was in this funny-looking spacesuit, crunched over in a little seat with two other astronauts in a tiny capsule. Nevertheless, it felt as good or better because it was real.

The first moment they told us that we could open our seat harnesses, I did, and I floated up to a porthole. I was so excited that I was giggling. I looked out and saw Earth, and I started to cry, because I felt this life energy and feeling of warmth come over me. I knew that it was in my head, because there was no heat source inside the capsule. I was just overwhelmed by the beauty of our home planet. One of my tears started floating in front of me, and when I saw that, I started to laugh. I was going between laughing and crying, and I thought, *My crewmates probably think I'm crazy and emotional*, and I was expecting them to say, 'That's why we don't want commercial passengers to go to space.' But they didn't. They understood.

[When I was on the ISS,] I started doing all sorts of things that I wasn't supposed to do, like somersaults and flying up and hitting my head on the hatch, because I didn't know how to move in microgravity. I applied too much pressure, thinking that I needed to push myself hard to move. I soon realised you need very little force to do anything in space. Then, after a few minutes, I got very sick, so I had to ask my crewmates to give me a shot to make me feel better, and that made me very drowsy for the next 48 hours.

I was told I could set up my sleeping bag anywhere I wanted, so I looked for the biggest window I could find and said, 'I want to be here.' Someone said, 'You're next to a fan there, so it's going to be noisy and cold. Are you sure?' I said, 'I'm absolutely sure.' Instead of sleeping, I spent most nights looking out the window. Every 90 minutes, you see a sunrise and a sunset. It's mesmerising. All my family, friends, schools, memories – everything was there, and here I was, looking at it. I felt a sense of safety, peace and freedom I had never experienced in my life.

I felt more connected to Earth being in space, and I felt immensely frustrated with humans. It was like I had been woken up from a big dream, and when I came back, all I wanted to do was shake people and say, 'Wake up. You can't see the reality of the world we are living in.' So I came back frustrated with humanity, but in love with our planet.

I was the first Iranian person in space, the first Muslim woman in space, and the first commercial space participant who was a woman. In the whole history of space exploration, there are probably only about 700 people who have gone to space, and only 10 per cent of those were women. So, compared to the eight billion people on this planet, only a very small fraction has experienced and seen space.

The response in Iran to me going to space was very positive, and people were very excited, especially young people. They started organising clubs and sightings of the ISS when I would fly over Iran. I started writing a blog, and people were sharing comments and saying how proud they were, talking about how me being able to achieve my dream had given them hope that they could achieve their own dreams. In particular, lots of young women started writing on my blog. After I came back,

Above: Iranians watch and take pictures of the ISS as it passes overhead carrying Iranian-born Anousheh Ansari on 23 September 2006.

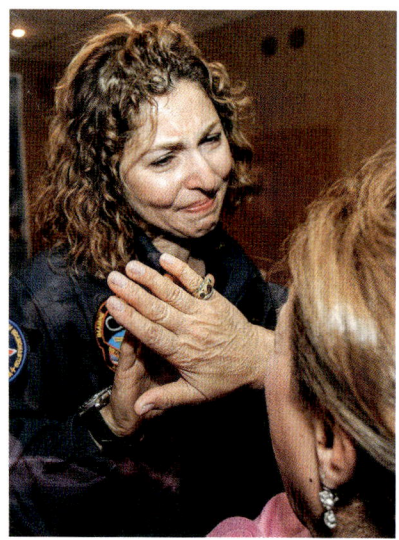

I found that 20 million people had been following and reading the blog and posting on it. It was overwhelmingly positive. A lot of people said how proud they were and how happy they were to see a woman succeed in something that usually, especially in our part of the world, is considered something that only men can do. I cried a lot reading many of them.

I never did any of this for publicity, or to be the first at anything, or for people to say that it made them feel proud or good about what I did. It was really more of a personal quest. But then seeing how I had been able to impact someone else's life positively was emotional. I remembered myself [being] in their shoes in Iran, a young girl, dreaming of something impossible. And I imagined how much it would have meant to me if I'd had someone who had given me hope. So, the fact that I was able to do that made me feel really good, but I also knew how difficult and long a journey the young women in Iran would have in front of them to pursue their own dreams.

There were also a lot of people writing to me on my blog to say that I [had] wasted a lot of money that could have gone to better use here on Earth versus spending it on going to space. The cost of going to space back then was about $20 million. However, I make a lot of philanthropic donations and support a lot of causes, and I also think the impact I had by going to space and demonstrating that it's possible for people like me to do this seemingly impossible task has inspired so many people and changed so many lives. A lot of my work in philanthropy is focused on eradicating root causes of the problem [rather than] trying to treat the symptoms of the problem. When I came back [from space], I realised the big impact this had on the lives of many young women and people who have dreams that seem impossible because they live in parts of the world that don't have access to a lot of resources. So, helping them through telling my story, talking to them, inspiring them, mentoring a few, became another big part of my journey.

Left: Anousheh's launch crew photograph for her Soyuz flight TMA-9.

Above: Anousheh Ansari says goodbye to her mother through a glass quarantine wall before leaving for space.

Opposite: Soyuz TMA-9 launches on 18 September 2006, carrying Anousheh Ansari on her pioneering flight to the ISS.

RICHARD GARRIOTT: In 2008, there was another market crash, so I went broke again the year I actually managed to get to space. Worse yet, once I'd paid all of my money to Russia, which I wasn't getting back, they did the medical pre-qualification and found I had a very concerning birth defect. Specifically, one lobe of my liver had no vein to drain it. On Earth, you wouldn't care [about that] unless you played a contact sport like football, but in space, if there was a depressurisation event, I would have an increased probability of internal bleeding, which would likely be fatal.

So, I received a phone call, having now sent all the money that I had in the world to Russia, to say, 'You've been scrubbed from the flight.' Fortunately, my US-based medical team jumped into action. They said, 'We're going to surgically remove that lobe of your liver, and the Russian medical board has agreed that if you can survive a 9G centrifuge run four weeks after surgery, you can start training.' So, that's what I did. I have a scar from my sternum to my belly button to the side of my body as one of the mementos of my space flight.

I became [Space Adventures'] sixth client, and I thought by then, NASA must be over this. I was also the son of an astronaut, and I was doing a ton of work for NASA on my flight. Yet NASA still tried to get me not to go. Even hours before launch, they were trying to change the rules and keep me from being able to enter the US segments of the ISS. Then, when I got back, if I shared any pictures that included a NASA astronaut and they made it onto the internet, I'd get a call from NASA legal to say, 'You have to figure out whoever put that up and get them to take it down.'

I consider myself a private astronaut and hate the term 'space tourist'. The dictionary definition of the word 'astronaut' is anyone who trains for or participates in space flight. So you become an astronaut even before you have flown, because you are training for space flight. But there was contention from the very beginning as to who gets to use the word 'astronaut'. As soon as we started flying private people, all the government astronauts said, 'Those aren't astronauts. We are the astronauts because it's our profession.' My reply to that was, 'You keep moving the goalposts.' NASA and the Russian space agency have a contract that says neither of those governments will use the word astronaut for us – the official term is 'space-flight participant', which I find kind of offensive.

Above: Richard Garriott's mission portrait.

Right: Richard Garriott with other crewmembers of Soyuz flight TMA-13 to the ISS.

SERGEI ZALYOTIN: If a non-professional has the chance to go to space, without breaching any safety protocols and the individual is properly prepared, then why not? The number of professionals who travel to space is very small. In Russia, on average, two out of 150 million people go into space each year, a tiny proportion. The same applies globally. Therefore, if you have the chance to fly without breaching any flight conditions or safety measures, and you are paying for it yourself, I see no issue with that.

THE NEW SPACE ERA

RICHARD GARRIOTT: The cost to send yourself to space has largely risen. When Tito went up, it was $20 million. By the time I went up, it was $30 million. By the time we sent Guy Laliberté, one of the founders of Cirque du Soleil, it was more like $40 million. Today, if you want to go to the International Space Station with a company called Axiom, it's about $70 million. So, that trend has been going in the wrong direction. But it's now starting to turn, and as soon as there's real competition, which is coming, the price should drop by an order of magnitude.

It is interesting that anyone who spends money on themselves of any significant amount will find people come out of the woodwork who are naysayers. All of our clients who have flown to space have had this happen. But while I did spend a heck of a lot of money going to space, I also did a lot of science while I was there, and I built other companies and found sponsorships to do experiments for other companies to help offset that cost. And I did things like buying carbon offsets for the fuel that we took onboard the vehicle. I also helped open a whole industry, so I feel very comfortable with my choice [to go to space].

I think fundamentally everyone understands that space is incredibly important and valuable to humanity. If you run the clock forward 50 or 100 years, we are going to need space-based resources to sustain this planet, and the only way that's going to happen is if we go through this period like we did with aeroplanes. The first people to put on their tux and top hat and fly across [the] country were similarly criticised as being wealthy yahoos, taking themselves on joy rides. But why was that a good thing? Because air travel is important, and so is space travel.

Every mature form of transportation – cars, boats, trains and planes – costs about three times the price of [the] energy [required] to operate it. So, if I put 100 bucks of gas in my car, I will probably have paid 300 bucks in total to cover its depreciation, insurance and maintenance. For rockets, that number is about a hundred to one, not three to one. But reusable rockets should get us close to three to one. We really are in this brand-new era, where the prices are already in the process of falling. And they're going to fall a lot more, which will unlock a revolution in what kind of activities and work and pleasure we can do in space.

WALT ANDERSON: During his presidency, Obama made the decision that NASA should be more open to New Space. There were a lot of people out there ready for that. Elon Musk benefited from that new policy. He was able to start SpaceX as a commercial operation and get NASA contracts. So, I'm happy that New Space is real now. I wish I had been part of it, but I was a little too early.

CARLOS GARDELLINI: For an orphan of Apollo, the dream ends on 11 December 1972 with the last Apollo landing. Then we have the [SpaceX] booster landing on 21 December 2015. Those are the brackets. We have been throwing away boosters since the invention of space [travel]. Elon lands one – game changer. Start the clock, we're all going. Why? Because we're now on the cusp of the airline model of space operations, which has been discussed my entire life. When you have a fully reusable vehicle, you fly for fuel. Up to now, the cost of space flight has been astronomical, because every time that 747 departs Los Angeles and lands in New York, you blow it up. When you stop blowing up the 747, you can fly coast to coast for $99. So we're going to reach a point, initially for the cost of a non-discounted, first-class transatlantic airline ticket, where we're all going to be able to be astronauts. I have no doubt of that.

RICK TUMLINSON: It's easy to get excited about what SpaceX or Blue Origin are doing now, because they're flying. When we were fighting this fight back in the late 1970s, 1980s and 1990s, we had no proof. We just had our belief. I am so honoured and privileged that I got to work with all of these people. I've watched people burn out over this. I've watched people lose their marriages over this. I've watched people lose their jobs over this. It has been a long, long fight. This is a revolution, and there are a lot of heroes, and they're not all billionaires.

In terms of opening the frontier, we haven't won yet by any means. There's a ton of work to be done, and everything could fall apart instantly. Elon Musk could pull his money and go somewhere else. Jeff Bezos could change his mind. The other companies could fail. God forbid, the government could put a restriction on the private sector opening space up. We will not have begun winning until there are people living out there [in space] who call it their home.

I think that we are at a moment in time where it could go either way. What's happening with the Earth's climate and the governmental response to that terrifies me. But then there's this beacon of hope with what's happening in space. And that excites me.

Right: SpaceX Starship SN15 and SN 16 at Starbase, Boca Chica, Texas.

Overleaf: The Moon rises behind a SpaceX Dragon capsule docked to the International Space Station in orbit above the Earth.

THE NEW SPACERS

PART FOUR:
THE NEW SPACE RACE

'Human space flight is a team sport. I love our international team.'

GINGER KERRICK,
NASA Flight Director

INTRODUCTION

On 20 November 1998, the first module of the International Space Station (ISS), built in Russia, launched into orbit. It was designed to meet and attach perfectly with a secondary module, then under construction thousands of miles away in the United States. These two modules had to fit together perfectly to create an airtight, enclosed habitat in which humans could live safely in the harsh environment of space. The first time both modules would be joined together was to be in orbit, 250 miles above the Earth's surface. The fact that they fitted perfectly was a marvel of engineering; that it required the cooperation of two former adversaries to make it happen was a miracle.

The ISS is the culmination of a long and evolving narrative of collaboration between America and Russia in space, one that began in the iciest depths of the Cold War. The 1975 Apollo-Soyuz Test Project, during which American astronauts and Soviet cosmonauts famously shook hands in orbit, was a symbolic moment of human connection between two countries that were poised on the brink of mutual nuclear annihilation. The handshake did little to ease the tensions on Earth, but it planted the seed of an idea that, in space at least, we might transcend political divisions. Almost a quarter of a century later, that idea culminated in the start of the construction of the ISS, a physical symbol for a more collaborative and peaceful future, both in the cosmos and on our home planet. Cooperation is built into the space station's DNA. While the Russian side handles guidance, navigation and control, which are crucial for orbit-keeping, the US Orbital Segment provides most of the station's power (from its large solar arrays), as well as a significant portion of the life support, thermal control, and air revitalisation systems. One side cannot operate properly without the other, making the ISS virtually impossible to control without the equal cooperation of both America and Russia. This is not a flaw in the design but rather its greatest strength; it necessitates constant communication, shared problem-solving, and a mutual reliance on each other to stay alive. The construction of the ISS is a tangible manifestation of mutually assured cooperation and a blueprint for how to explore the solar system as a unified species with a shared curiosity. In an era often marked by division, the image of a multinational crew living and working together in space offers a powerful and inspiring counter-narrative … until recently.

In 2022, Russia invaded Ukraine, bringing war, once again, to Europe. Once again, Russia and America found themselves on opposing sides. Sanctions were enforced, trade was disrupted, business ties were severed, and multinational companies closed their Russian operations. Nearly all cooperative endeavours between the West and Russia were stopped. There was only one area in which America and Russia continued to cooperate, and that was in space, on the ISS. As Ginger Kerrick, who trained the first astronauts to inhabit the ISS, observed, it is a sign of the remarkable ability of space to overcome earthly politics. As the relationship between the two countries deteriorated after the invasion, the only meaningful cooperation to survive between the two was on the ISS – a testament to the important regard in which both governments held the project. However, that does not mean it was business as usual. Cooperation on the ISS may have continued, but the relationship in space was stress-tested like never before. And as the war intensified, the politics of the situation slowly crept into space.

In this chapter, we hear from those astronauts and cosmonauts who witnessed the political contamination of the space station. From threats to maroon astronauts in space to the unfurling of pro-Russian flags, it became apparent that politics could no longer be confined to Earth, and that the long history of Russia and America working together in space looked set to end. Russia and America are now no longer the only players in this game. India, China, Saudi Arabia, Israel and Japan have all joined the race, and this time it's not about a symbolic flag on the Moon to prove ideological superiority.

This new race is driven by ambitions to establish a sustained human presence in space and gain access to potentially valuable resources. Securing these locations could promise huge wealth and power back here on Earth. Governments and billionaires are once again reaching back out to space. Jean-Patrice Keka, a Congolese rocket engineer, is among the less likely players fighting for his seat at the table. Working on a shoestring budget, with rockets made largely from repurposed materials, he is on the verge of sending the first African-manufactured rocket to space, launched from African soil. If he succeeds, he wants all of Africa to have a claim in space, to share in the benefits that spacefaring nations are reaching for. History has taught him all too well what it means to be shut out of the opportunities that new frontiers offer. This time, he wants Africa to have a stake in the future.

Previous page: View of Earth through the windows of the Italian-built Cupola on the Tranquility module (Node 3) of the ISS.

Overleaf: Set against the backdrop of the deep blackness of space, the ISS is seen in orbit from the Space Shuttle Discovery.

By 1996, the Shuttle-Mir programme had successfully delivered 21 American astronauts to the Russian space station. Their collaborative efforts were proof that the working relationship required for the larger, more complex International Space Station (ISS) was possible, and development and planning of the ISS could begin in earnest. To shoulder the monumental task of assembling the station in orbit, in the summer of 1996 NASA hired their largest cohort of astronauts to date. Nicknamed 'the Sardines', the 44 new astronauts would be instrumental in the construction of the ISS. Ginger Kerrick was one of the applicants.

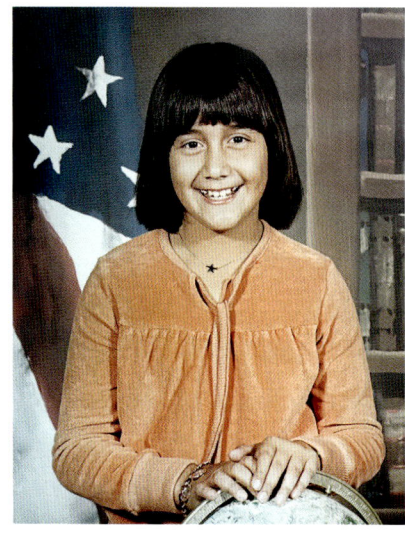

GINGER KERRICK: When I was five years old, I used to go to the library with my parents every Friday afternoon, and I remember being really excited about this book called *Astronomy and Astronauts*. There was a passage about what it felt like when the astronauts stepped outside and looked at the Earth, and I thought, *I want to see that*. That feeling stayed with me, and my singular focus was on being an astronaut. I didn't even really investigate what other careers were available at NASA. The minimum qualifications to be an astronaut back then were a master's degree and one year of technical experience. So, when I got to NASA in May 1994 with a master's degree, I basically set a timer. Then, in May '95, I filled out the form and hand delivered it to the astronaut selection office.

When I went to turn in my application, Duane Ross [head of the astronaut selection board] asked if I had ever considered working in mission operations as an astronaut instructor. I said, 'I didn't even know that was a job!' And he said, 'I think you might be good at it.'

While waiting to find out if her astronaut application had been successful, Kerrick began working as an instructor, developing training materials for the ISS.

Three thousand people applied [to the 1996 class of astronauts], and I was one of 120 that they wanted to interview – I lost it. I called everybody. I even called my old school who said, 'Astronaut? No one from El Paso is ever going to be an astronaut.' It finally came down to a one-hour interview, and psychological, medical and physical fitness tests. During the medical tests, they were doing an ultrasound of my lower abdomen, and the technician said, 'Ma'am, I'm going to need you to come in tomorrow morning for a CT scan.' I had kidney stones. Back then, if your body showed the ability to form even a single stone, it was a lifetime disqualification. I don't remember the rest of that day, but I remember being home that weekend and just crying.

It was at that moment that I thought, *No, no, no. This is not me. I am strong. I am fierce. I can handle anything*. I took some deep breaths and analysed the situation: *I can never go into space, but if I continue teaching the astronauts, each one of them can take a piece of me with them*. I was open to doing the best I possibly could in that role and seeing where it might lead. Now, I always tell people that I'm kind of glad I had kidney stones. I would have loved to have been an astronaut – they have a much better view than I do from the ground at NASA – but, in reality, the career I had [supporting the crews in various guises] was the one I was supposed to have. I think I was able to leave NASA better than it was when I found it, and that's the definition of success.

Left: Ginger Kerrick as a young girl.

Above: NASA's astronaut class of 1996.

Right: NASA Astronaut Group 16 Patch: representing the largest group in the history of the US space program. On the patch, the Space Shuttle and ISS above Earth along side the Moon and Mars represent the challenges of their work ahead. The infinity symbol represents the never ending quest of human exploration and the eight flags bordering the patch represents the international nature of the group.

THE NEW SPACE RACE 249

Dan Tani was one of the successful candidates chosen in NASA Astronaut Group 16.

DAN TANI: I grew up hearing stories about my parents being in the [internment] camp. Some of those stories were of hardship, but some were just of general interest. I don't remember my mom ever sitting me down and explaining that during World War Two anyone of Japanese heritage was pulled from their homes, but I knew that was part of our family history. After Japan attacked Pearl Harbor in December 1941, many Americans believed that the Japanese living in America couldn't be trusted, that they were spying for the enemy. So there was a presidential executive order to remove anybody of Japanese ancestry from the West Coast: 120,000 people were relocated.

My parents were in their thirties. My mom [Rose] was born just outside of Sacramento and my father [Henry] in San Francisco, so they were born US citizens. They were married and had a three-week-old son, my brother [Dick]. But they were put on a train and sent out to Utah, where they lived behind barbed wire in hastily made barracks in the middle of the desert for the next two years, with armed guards and limited access largely to the outside world. There was no bitterness. I would ask my mom, 'Did you not rebel against this?' and she would say, 'This is what the government were asking us to do. This is what we did to be patriotic.'

Following the death of Tani's father Henry when he was four years old, Tani was brought up by his mother Rose, but unlike many astronauts, his sights weren't set on space from a young age. After high school, he went to MIT to study mechanical engineering.

I was eight years old when Neil and Buzz landed on the Moon [in 1969]. Everybody was space and astronaut crazy at that time, but it certainly wasn't something that I thought was on the horizon when I was growing up. My journey to space was serendipity, not a master plan. One of the games you play as a senior is, 'Where can I visit for free?' So, I looked at all the companies that were interviewing, and one of them was an aerospace company out in California. I thought, *I'll go talk to them, because if I can get a free trip to Los Angeles, that'd be kind of fun.*

It was snowing in Boston, and it was 75 degrees Fahrenheit [24°C] in Los Angeles when I landed. I interviewed, and [I thought] it was cool that they were making these big satellites. I'd honestly never really thought much about space until then, but I thought it would be silly to turn this down – from a lifestyle point of view, it was going to be a ton of fun. All of a sudden, I was living near the beach and goofing around in the design department of this aerospace company.

After doing a master's at MIT and working for the research and development company Bolt, Beranek and Newman, Tani joined aerospace company Orbital Sciences Corporation in LA in 1988. He applied to NASA in 1995.

One day somebody said, 'There's a notice up that NASA has called for applications for the next astronaut class.' And I did what you do when you're in your early thirties – I laughed and said, 'Wouldn't that be cool?' I wouldn't have bet a dollar that my application would go anywhere. Finding out I had an interview was like a thunderbolt – unbelievable. That's when my imagination started going crazy: Oh my God, this could actually happen. It became something I really wanted. It was like I was being offered a knighthood and there was absolutely no question that if I had a chance to be an astronaut, then that's exactly what I'd want to do.

I was selected in 1996, and NASA had not flown any element of the International Space Station yet, but they were getting ready to start building it. We were looking at four or five flights a year. To sustain that flight rate, you need 100 to 120 astronauts to constantly be in training and in readiness to fly. It was great news for us that they needed a big class because they anticipated a lot of flights and a lot of space walks – that was just music to our ears.

Willie McCool was also selected as part of NASA Astronaut Group 16. Initially, his teenage son Sean was not impressed that his dad was going to be an astronaut.

SEAN MCCOOL: My dad was a pilot in the navy, and after that he was a test pilot. He was top of his test-pilot class, and everyone there was applying for NASA, so he applied too. He was on a ship doing training when he got accepted in 1996. I was a too-cool-for-school teenager, and space flight had become routine, so I didn't even think it was that big a deal.

Houston was all about the military and 'Go America'. That was a culture shock for me, and I didn't want to move there. I didn't bring up that my dad was an astronaut, but when I made friends and they eventually found out, they thought it was very cool. I also got a job working at the [Johnson] Space Center in Houston as a tour guide, which showed me that it really was a big deal.

My mom and dad got together when I was six, but he never referred to us as his stepkids – he kissed us goodnight, and even when we were adults, he would still kiss us goodbye on the cheek. I didn't realise that I was so lucky to have a stepdad who just saw us as his own kids. The word 'chivalrous' is not used to describe many people these days, but that was what he was: doing the right thing with honour, and doing the right thing even when no one was looking.

Left, above: Dan Tani's parents and brother, Rose, Henry and Dick, in Topaz internment camp, 1942-1944.

Left, below: Dan Tani with his mother, Rose, 1965.

Right: Future NASA astronaut Willie McCool and his son Sean kayaking in the 1990s.

In 1993, Vice-President Al Gore and Russian Prime Minister Viktor Chernomyrdin announced that the USA and Russia were going to collaborate on a joint space station, along with the European, Japanese and Canadian space agencies. The construction of the first module of the ISS began on Earth in 1994, ready for assembly in space.

GINGER KERRICK: When I became an astronaut instructor, the first crew to fly on board the ISS had not yet been assigned. Then in May 1997 I began my first class, teaching the Expedition 1 crew. They were going to stay up there for about four to six months. The crew was made up of Captain Bill 'Shep' Shepherd, commander of the expedition, Sergei Krikalev, flight engineer, and Yuri Gidzenko, flight engineer number two and commander of the Soyuz spacecraft.

Life changed considerably with the responsibility [of training astronauts to go to the ISS]. I was living in an apartment at the time, and I tried to keep thinking that I was only going out to Russia for a trip or two, but I quickly realised that I needed to live in Russia, because the crew were there.

I loved working in Russia. I lived in a dormitory that Yuri Gagarin had actually stayed in. The people in Star City were also really inviting. It was a secret military base, and they hadn't worked with a lot of female American engineers, so I was a novelty for them. They would open doors and carry my bags and make sure I had tea and snacks – everybody was really friendly. It was wonderful.

It didn't at all feel like I was going into the territory of America's enemy. Before my first trip, my briefing said, 'Don't

wear bright colours. Don't look anyone in the eye. Don't talk to anybody on the subway. Walk fast. Be careful what you eat, but don't refuse any drinks, because that's a sign of disrespect.' They load you up to be completely paranoid, but I didn't see a reason to be suspicious or nervous. They were just people, and most of them had something in common with me: a passion for exploration and human space flight. If you start there, it allows you to integrate a lot more quickly than if you go into it with preconceived notions.

We spent a lot of time together – me, Yuri, Shep, Sergei and my colleague Chris Niemann, with whom I split the systems that we were working on. We would all go grocery shopping together and to the embassy to work out. And then we had everybody who was working on the American side at Star City and our Russian counterparts supporting the expedition – all of us were one family.

Above: Ginger Kerrick, 2016

Below: Ginger Kerrick with the ISS's first crew, Expedition 1, before their flight in 2000.

Right: Ginger Kerrick, wrapped up on a winter visit to Moscow in the early 2000s, in front of St Basil's Cathedral in Red Square.

On 4 December 1998, STS-88, the first space shuttle dedicated to assembling the ISS, launched into orbit.

GINGER KERRICK: The first piece of the space station flew up in November 1998, and that was the FGB, the functional cargo block, which was manufactured in Russia.[1] Our Russian counterparts had sent up the FGB by themselves. STS-88 was the first time the Shuttle had been used to carry an ISS module into orbit. Led by Bob Cabana and Sergei Krikalev, it took the first American module, Unity, to the station. It was what we call an outfitting mission, where you go inside the module and install some hardware. It was the first time anybody had gone into these modules since they had been on the ground, and it was the first test of us operating the space station together.

Historically, the commander of the shuttle goes through the hatchway first, but Bob recognised the significance of the moment. This was an international space station, so he said to Sergei, 'When I open this hatch, we're both going to go in together.' I burst into tears and thought, *How beautiful is that*. It's *our* space station. It was just as it should be. If you're going to work together, then you should be doing things that symbolise your unified approach to operating and living on board the space station.

We had a joint party with the Americans and Russians, and there was toast after toast celebrating [the launch of the FGB], because that meant our adventure was beginning. I was just happy to have been a part of it. I thought, *I didn't get to go to space, but I've been inside that module, and it went to space.* It was something that I was a part of, so it was a part of me. I felt pride, and a little ownership. I didn't look at it and think, *This is a Russian vehicle.* I thought, *This is our joint spacecraft, and our first step in the ISS.* And my Russian counterparts cheered when our lab module launched and was attached. We really did cheer for each other collectively.

Cooperation is part of the DNA of the station. The space station was designed so that neither the USA nor Russia can operate the vehicle independently of one another, and it was intended to be able to withstand any political turmoil between our countries at any given time. For example, the USA keeps the power running, and our Russian counterparts make sure that we can stay in orbit with the right propulsion. The ISS symbolises a truly unified approach, and this has helped us to survive a number of challenges, with respect to the political situation. I love that about it. If we'd had suspicions about one another, we wouldn't have selected a design that implied we'd be able to figure out how to bring the best out of each other and how to work together for the benefit of this wonderful platform that we'd created.

It was a partnership of opportunity that allowed the USA to continue our dream of having a space station, and we recognised that it was easier to fund if there were multiple collaborators. I don't think people realised what it was going to do to the humans who worked on it in terms of the camaraderie that we felt. That's why you hear a lot of astronauts who come back and say that we've got to do better in how we exist on this planet and how we treat each other, because if we can accomplish this massive undertaking, this engineering marvel that is the space station, by working together, what else can we accomplish?

---

[1] The acronym 'FGB' relates to the Russian term for the functional cargo block, also known as Zarya.

Right: The first two modules of the International Space Station are joined by the crew of Space Shuttle Endeavour. Unity sits in the payload bay of the Shuttle, whilst the Shuttle's robotic arm is used to manoeuvre the Russian module, Zarya, into position, 6 December 1998.

Cosmonaut Sergei Zalyotin was part of the final resident crew to visit the Mir station, before it was deorbited in 2001. Two years after his return to Earth, he commanded a Soyuz mission to the recently launched ISS.

SERGEI ZALYOTIN: The ISS was the logical conclusion of our first interactions in space: The Soyuz–Apollo and the Shuttle–Mir programme. The first modules [of the ISS] were launched while Mir was still operational. While we understood that the ISS was the next step forward, in our hearts we knew that we were also slowly winding down our own manned space programme. However, the vast majority of cosmonauts and citizens of our country saw this positively. Anything that would bring Russia and America closer together and was positive for relations between the two countries was considered to be a good thing.

At the time of the first launch, I was working at the Johnson Space Center in Houston, where I was overseeing training and helping them to prepare – in particular, the first crew of Gidzenko, Krikalev and Shepherd. This was one of the best-trained crews that had been formed during my entire time working in the space programme.

At the beginning, when we simply didn't know each other yet, we thought that the Americans always had their own special thing going on and were different to us. But when we sat down together, had some tea or drank a glass of cognac, say, and talked, we saw that we had the same attitudes. It was the beginning of a more trusting, warmer relationship.

On the International Space Station, we breathe the same air, we drink the same water. Everything is communal. It is a joint construction, the highest technical achievement made by all those countries that participated in the programme. But to be honest I didn't really like living on the ISS because of its large size. I always preferred Mir.

Below: A historic handshake in space between American astronaut Thomas Stafford and Soviet cosmonaut Alexei Leonov during the Apollo-Soyuz Test Project, July 1975.

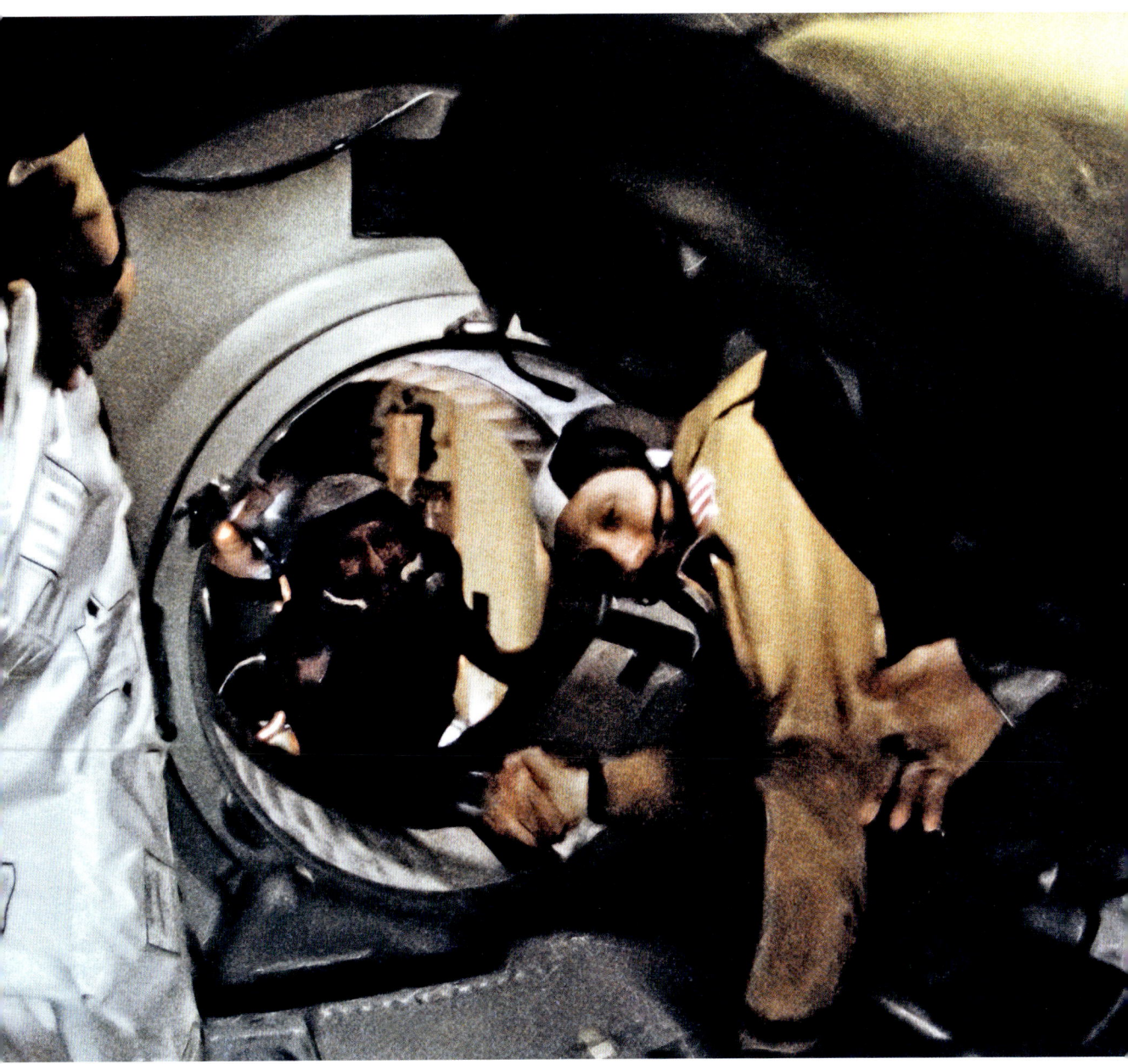

Terry Virts was selected in NASA Astronaut Group 18 in 2000, recruited to expand the ISS contruction effort.

TERRY VIRTS: I first came to NASA in 2000 when we were building the space station. Back then, it was all about getting the station built; we had to get a certain number of the primary modules built by April 2004, called 'core complete', so there was this big push to launch all these missions. The excitement about the ISS was really palpable. Everybody felt it. It was a time when we were friends with the Russians. We used to say, 'Off the Earth for the Earth.' There was this excitement that this was how people could and should work together: 'We have all these problems down here on Earth, but let's build this space station.'

I loved the International Space Station – I was a true believer. I loved that it was a way for us to work together, but it was also an engineering achievement to put this thing together in space. It was really impressive. There's all kinds of experiments that have been done there, but for me, the most important part of the space station was the international relations. I think it [has been] the most successful American foreign relations initiative since the Marshall Plan and the rebuilding of Europe after World War Two.

Designing and building the space station was an amazing achievement partly because the Russian approach is very different. For example, we had this complicated water recycling system; the Russians had a tank of water with a pump and a rubber hose that you could drink from. So, they had very simple systems that worked, and we had complicated systems that could do a lot more but were expensive and prone to breaking. There is an advantage in keeping things simple – sometimes that's the right solution. But on the other hand, the Russians didn't really innovate. They came up with the designs in the 1960s, and they haven't changed since then. It was interesting to see these different design philosophies at play, and it was a good marriage. You want some differences – that's what attracts us in a relationship – and it worked well in the space station.

On the ISS there's a Russian segment and an American segment, and the American segment really means America plus Europe, Japan and Canada. They use different voltages and data protocols, and the hatches are different sizes. It's very obvious that these are two different places, but we all share the same air and water. So, even though there are different types of modules, the things you need for life are shared.

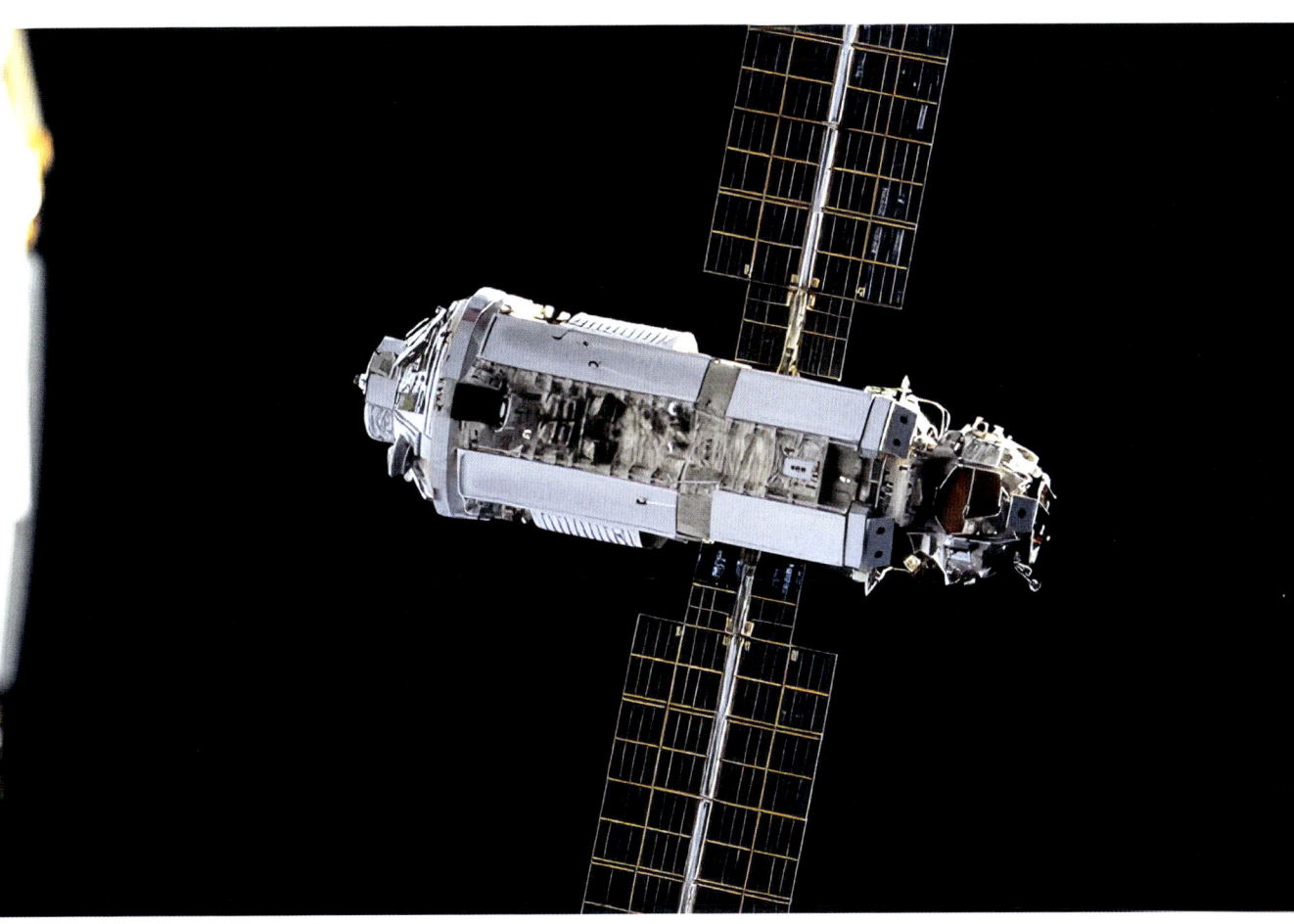

Left, above: NASA astronaut Terry Virts.

Left, below: Virts gives the *Star Trek* Vulcan salute to Boston, Massachusetts — Leonard Nimoy's hometown — after hearing the news of Nimoy's death in February 2015, following a long career playing the character Spock in the famous science fiction franchise.

Above: Zarya, the first module of the ISS to be launched, on 20 November 1998.

EXPEDITION 1

While the first modules of the ISS were launched into orbit, astronaut trainer Ginger Kerrick was busy preparing astronauts and cosmonauts for Expedition 1, the first long-duration crewed mission to the station, which launched on 31 October 2000.

GINGER KERRICK: With any [space] programme, whether it be Gemini or Apollo, people always remember the first mission, and it always has a little bit more risk to it than any other because you're just learning how to operate the spacecraft. We knew that we wanted to keep the ISS up for 15 years at that time, so we needed to get this right. This crew had to install the life-support equipment, so we were flying them up to a space station that didn't even have the equipment that would keep them alive activated yet. The Expedition 1 crew set the bar when they launched in 2000 and started a clock of continuous human presence in space on board the space station.

As the days got closer [to the launch], I felt excitement combined with apprehension. All the questions hit me: Did I give everything? Do we have everything? It had been three years of my life trying to get these guys ready, and I wasn't going to be able to meet them tomorrow to go over things again, since they would be in space and my job would be done. I just wanted to make sure that they had everything they needed, and that feeling never went away.

[Before the launch] the crew always go up the ladder and then turn back and do their standard pose for a photo. I watched them do that on the big screen, then they went up the stairs

and disappeared [through the hatch]. It was right around then when I couldn't see them anymore that I started to hyperventilate. A gentleman came up who had seen me with the crew before, and said, 'It's OK. Come with me.' He began to part the crowd – it was really tightly packed – and said in Russian, 'It's the mother of the crew. It's the mother of the crew.'

He helped me to fight my way up to the front, where there was a smaller TV screen on which I could now see inside the capsule. All of a sudden, I was OK again, but I felt like an idiot. I knew that they were trained and that they were ready, but I wasn't ready. I had spent so much time focusing on them that I hadn't realised what this whole experience was going to do to me, and I was terrified, because it was a rocket and they were going to be on top of it. Eventually, I let my physics brain tell me that everything was OK.

I was just so proud of them. This was the beginning of it all, and it was bigger than me, and it was bigger than the three people launching on that cold, foggy day in 2000. Back then, I just saw three members of my closest family going into space, so I cried for a different reason. Now I cry for what they set in motion and how beautiful it is.

Left: Expedition 1 crew members pose for final photos prior to their launch aboard a Soyuz vehicle from the Baikonur Cosmodrome in Kazakhstan. Commander William M. (Bill) Shepherd (centre) is flanked by Soyuz commander Yuri P. Gidzenko (bottom) and Sergei K. Krikalev, flight engineer (top).

Below: Expedition 1 at Baikonur launch site as the preparations continue for the launch of the Progress M-66 cargo vehicle under the International Space Station programme, 2009.

GALAXIONAUTS

Left: The Keka family.

Right: Africa is front and center in this image of Earth taken in July 2015 by a NASA camera on the Deep Space Climate Observatory (DSCOVR) satellite. The image, shot one million miles from Earth, was one of the first taken by NASA's Earth Polychromatic Imaging Camera (EPIC).

Jean-Patrice Keka watched with interest the development of the ISS from the Democratic Republic of the Congo, where he was already on his way to becoming a pioneer of space development in Africa.

JEAN-PATRICE KEKA: All my efforts started to converge from 2000 onwards, around the time the ISS became operational and all the greatest minds working in aerospace were coming together to find solutions and create this remarkable international spacecraft. But it was only international to the extent that it let all the capable countries with expertise in the field participate. There was a missing piece in the puzzle: Africa. We missed out because of our lack of expertise. And that's why I'm taking this opportunity to tell all Africans that we need to get to work, and there needs to be an awakening, because we need to contribute to the evolution of science that allows us to understand the universe. The universe is for everyone. We can't be the missing link in the chain. We absolutely need to be part of it.

The idea of creating an international space station had a great impact on my life and space research, because I really believed in it, and it made me understand that space research is international, meaning that even I could do it in my corner of the world and the Democratic Republic of the Congo (DRC) could contribute. It greatly motivated not just me but a lot of people who saw something being built across the world that any capable person could contribute to. And that was extraordinary.

My big dream is to see a rocket that I built be the first African rocket in space. And this rocket could transport African satellites and send Africans to space too. I've even given them a name: America has astronauts, Russia has cosmonauts, China has taikonauts, Africa will have 'galaxionauts'. This would allow African youths to be on the same scientific footing as all other young people. If I achieve my goals, I can die a happy man.

I was born in a village in Katako-Kombe, a rural area of the DRC. My father was a teacher, so he encouraged us to study, and my mother, who was very affectionate, was always there taking care of us. Some of my elder brothers and sisters, who had gone to university and got married, moved to big cities like Kinshasa, and they brought us little ones there to live with them. So, I started my life in the village up until about the end of primary school, and everything else after that was away from home. But I was born a villager, and I am proud of that, because that's what made me who I am.

Space was a childhood dream, because, when I was still very young, I was always building micro-rockets. I had no mathematical knowledge,

no knowledge of physics, no chemistry knowledge, but my rockets would launch. And then that dream grew. In fact, my appetite for it grew to such an extent that at some point I thought, *Couldn't I go to space myself?*

It wasn't just rockets, though. As a young man, I was captivated by all things scientific. Every little gadget I came across would interest me, and I'd cobble things together. I had a friend whose father was the vice-governor of the Eastern Province in Kisangani. Each time I'd go to his house, we'd watch films on his projector. I thought, *Why can't I have that?* So, I built my own device, and my friend gave me films, and in the evening the whole family would gather around my device and watch. People were saying, 'He's a genius. We've got to support him.' I even built my first glider using fibres from palm trees and materials from the forest, and it worked.

When I was 11, I had to go back to Kinshasa to continue my studies, but I didn't want to leave all my lab equipment behind. I was travelling on a big lorry, with a lot of passengers, and on the way, some sulphuric acid spilled out of my case and fell on the steel floor, creating a chemical reaction.

There was smoke, and people started to scream, 'Fire!' The driver stopped the vehicle and said, 'You have to make a decision now. Either we leave you here with your suitcase, or you leave your suitcase behind.' If I'd stayed behind, a feral beast might have eaten me, so I had to leave my suitcase there. I cried for the entire journey.

When I was about 17, I used matchsticks to create a rocket that was bigger than my previous ones. I couldn't launch it from my plot of land. The only good place was in the streets, and there was a power cut that evening, so I thought I'd take advantage of the lack of people outside. When I launched it, there was a smoke trail and a hiss, and people were able to see it from afar. The police arrived and I was arrested. In the morning, when their chief arrived, he said, 'Why did you arrest him? Can't you see he's a scientist?' and he brought me to the Ministry of Defence to put me in contact with experts and engineers who had studied all over the world. They welcomed me, shared their research materials and started training me when they saw what I'd done. That was my introduction to rocket science theory.

LIFE ON THE INTERNATIONAL SPACE STATION

The goals of the ISS were far-ranging and ambitious. As well as being an orbiting laboratory, the purpose of the space station was to facilitate long-duration space flights that would help prepare humans for deep space exploration. Four-to-six-month stays on the station therefore became the new norm.

GINGER KERRICK: After the Expedition 1 crew launched, I spent some time at Russian Mission Control, and then I eventually moved back to the USA and took some time to reassess what was next for me. I wanted to find the way in which I could best contribute, so I went to talk to my boss at the time, Randy Stone. He said that a really good place for me would be in the control room as a CapCom, short for 'capsule communicator', the person who talks to the crew.

I said, 'Randy, I can't be a CapCom. CapComs are always astronauts.' He said, 'Well, do you know why they've always been astronauts? The people up in space wanted to talk to somebody on the ground who understood the vehicle. This is a brand-new space station, and you're one of the few people who knows all about it, so why don't you give it a try?'

I was honoured, of course, to be the first one, but there's also a lot of pressure any time you're the first to do something. I kept telling myself, 'Don't screw this up, Ginger.' On my first day on the console, I was so nervous. I walked in and introduced myself to the flight director, then the crew called down. That was my first time talking to space. I heard someone say, 'Ginger, is that you? Congratulations! I heard this was your first day, and you sound really professional.' Then a Russian voice said, 'Gin, is it really you? It is like a present to hear your voice from mission control – Misha, come, it's our Gin!' It was Vladimir Dezhurov and Mikhail 'Misha' Tyurin, who I'd trained, so it was a really great first day.

Because I'd spent time with the crew, I would sometimes hear conversations on the ground and say, 'We better tell the crew this. And we better tell them this way.' Similarly, the crew would call down with requests that sometimes the ground didn't fully understand, because the crew was on the scene with the actual hardware. Since I understood their training and knew them so well, I could interpret what they meant. So, it's like you're the interpreter between the crew and the ground.

I always thought it was my job to keep it light and keep them entertained – these folks are up there for four to six months. We would keep it business focused, of course, but still personal. Everything keeps going [while you're in space], so we tried to do our best to make sure the crews could follow along with us. We sent up Christmas trees, flags for the Fourth of July, costumes for Halloween. I think the one thing we learned about staying on the space station is that it's a marathon not a sprint, so let's make sure that we take care of our astronauts. [On the IP phone] they can call anybody in the world at any time, and once a week they can video conference with their families or friends back home.

Before the launch, there are conversations between the chief of the astronaut office and the crew members, and we ask them if they want to be told if a close family member passes away. There have been a few who have elected not to know, but the majority do want to be told. It's a scenario you can go over in your head before you fly, and a lot of astronauts have an amazing ability to compartmentalise, but sometimes your plan doesn't match your natural emotional response.

Below: inside NASA's Mission Operations Control Room in Houston.

THE NEW SPACE RACE

DAN TANI: On the space station, we are operational from about 8am until about 8pm, and very rarely will you get a call after 8pm. It was after 10pm, and I was talking to [my wife] Jane about the day. On the communication system, CapCom said, 'Station Houston, on space to ground one' – that's the communication channel – 'we're going to set up a private call on space to ground two for Dan.' So, just that happening was like the phone ringing in the middle of the night and you know it's something serious. I said to Jane, 'Holy cow. They just gave me a call for a private conference. Do you know what's going on?' She didn't, but they weren't going to call you on a private line because you'd won the lottery or something. Nothing good was going to come out of that call.

We hung up and then I was waiting for 20 or 25 minutes. Once you tell me you're going to set up a private call, I know it's bad news, so I want to hear it right away. Finally, the call got set up, and I said, 'Just tell me what's going on. Just give me the basics.' And Jane, who'd come into mission control, got on and

266 THE NEW SPACE RACE

Left, above: Astronaut Dan Tani, 20 August 2008.

Left, below: Dan Tani on a space walk, 29 June 2007.

Below: Dan's mother, Rose Tani, on the flight deck of a shuttle simulator in 2001.

said, 'It's your mom. She was killed in a car accident.'

It was emotional. Nobody gets training for losing your mother, but you deal with it the best you can. I would be on the phone for hours, talking to my family, talking to Jane, and that would make me feel better, helping to make arrangements, the way families do at a time of crisis. But I was unable to participate in any of the hugs or physical consoling. It was impossible for me to get home.

It was a day or two later, and there was a moment of quiet time when I wept because I wasn't going to be able to talk to my mom anymore. The enormity of it all hit me, and it was good to have that time to really mourn her.

I wanted to be part of the funeral, so I created a video [in space] they could use in the service. [In it, I said,] 'I have feared this moment for most of my adult life. Of course, I never suspected that I'd have to speak on this occasion on videotape . . . My mom loved and embraced life, its ups and its downs. And I will do better to live in her model, to accept life's triumphs and tragedies, including this one, and continue to live life to its fullest. Mom, you are a wonderful inspiration to me, and I will try to honour you by passing on your spirit, love and joy to all of those who are important to me, most of all my children. Mom, I love you so much.'

We have the International Space Station because we want to learn what it's like to live off this planet, but if you're going to live off the planet, all aspects of life are still going to happen. So there's going to be wonderful things and there's going to be terrible things. I wrote down my lessons learned and left them on the space station. I don't know where that ended up, but I wanted to leave some advice for when it happened to the next person.

GINGER KERRICK: It's not all roses down here, and it's the same up there. These are real people with real lives who have chosen a noble profession to advance humanity, and we have to remember that they are not robots. We have to take care of them while they're up there. Space flight is important, and the work that they do is important, but our astronauts are too.

THE NEW SPACE RACE 267

THE COLUMBIA DISASTER

Below: The crew of the STS-107 strike a 'flying' pose in the SPACEHAB Research Double Module (RDM) aboard the Space Shuttle Columbia. Top row, in blue from left to right: mission specialist David M. Brown, pilot William C. McCool, and Michael P. Anderson, payload commander; Bottom row, in red shirts from left to right: mission specialist Kalpana Chawla, mission commander Rick D. Husband, mission specialist Laurel B. Clark and payload specialist Ilan Ramon representing the Israeli Space Agency.

NASA's fleet of shuttles, in tandem with Russian Soyuz rockets, was instrumental in building the ISS, ferrying crew up and down and delivering supplies. However, with construction of the ISS still underway, its participation was paused when NASA's Shuttle Columbia disintegrated during re-entry of mission STS-107 on 1 February 2003, killing pilot Willie McCool and the six other crew members.

SEAN MCCOOL: When my dad eventually flew, it had been delayed so long and built up so much, I thought, *Right, this is finally happening.* The immediate family members watched [lift off] from the rooftop of the launch control building. There were astronaut escorts there, and they were describing what was happening over loudspeakers. The clock was already counting down, even though it was a couple of hours away. In fact, that clock had been counting down for a couple of days. It takes that long to prep for everything.

[I said goodbye to him in person] at the beach house. I also got to speak to him on the phone the morning of the launch. I remember being nervous, and I told him I was proud of him. He just laughed and said, 'I'm proud of *you*. Good job in college. Say hi to your friends for me.' That was my last time to talk to him. I remember thinking, *I don't know if people tell their dads that they're proud of them, but on the day of the launch I really felt, Wow. I'm really proud of you.* I'm glad I got to tell him that, though of course he turned it around. It's a good memory.

[At lift-off,] there's this huge delay, then the sound hits you, and it feels like a deep, rumbling bass inside your chest, even though you're miles away. It's just that powerful. After two and a half minutes, the two solid rocket boosters break [off], and that's when everyone starts cheering really loud at every shuttle launch, because that was what took out Challenger. That was the most stressful part for me, waiting for the solid rocket boosters to break off. When they did, everyone assumed they were in the clear – they'd made it past the dangerous part. It was like a big collective sigh of relief, and you could feel the tension lift, not just for me, but for everyone there. It was very emotional and really intense.

[After the launch,] I went back to school in Quebec, because I'd already missed school, and I couldn't afford to miss more for the landing. And the landing wasn't really a big deal. The launch was the big deal; the landing was routine. We thought after the launch, everything was going to be OK.

The internet was really slow when I was watching the video [of re-entry]. I could only get the audio – the video was just frozen – but after a while, I could tell something wasn't right. It was,

'Columbia, Houston, radio check,' over and over, and then eventually I heard the flight director say to lock the doors and do something about emergency procedures. I was still wondering what was going on, so I called Dan Tani, who said, 'Sean, where are you? Doesn't look good.' He was very blunt, which was what I needed. By the time we drove back to my girlfriend's house, it was all over the news. I called my mom, and she was in a panic.

It was a long, sad plane ride home. It was surreal. You fall asleep, and when you wake up, you've forgotten about it. Then you get home and see your mom and your brothers, and you have to relive it all. Unless you've experienced a shocking, unexpected death of a loved one, or something really big like that, it's hard to explain.

Dan and Jane Tani were there for us. We couldn't have asked for better care and better friends. I'm sure it was really hard not being emotional in front of us, because then that would make us more emotional. There were memorial services in DC and in Annapolis, where he went to Naval Academy, and I was able to hang out with Dan and crack jokes, and I really appreciated his friendship and having someone like that. The Tanis will always be special [to me].

I feel guilty now for not doing a better job of taking care of my grieving process. Whatever the steps are supposed to be, I pushed them all away and went back to school and then joined the marines. It was too painful to deal with, so I didn't. But you can't run away from it for ever. My wife was telling me, 'The kids should know about him. People should know about him. He was such an amazing person. Not just because he was an astronaut, but just as a dad and as a human.' It's something I've been working on these past few years, but it's still hard, and it'll always be hard, especially around January and the anniversary of the launch. That almost feels as sad [as the landing date], because once they launched, there was no coming back. It was a slow-motion accident waiting to happen, and we just didn't know it.

Above: Space Shuttle Columbia rolling to Launch Pad 39A atop the Mobile Launcher Platform with the crawler-transporter underneath.

THE BIRTH OF AN AFRICAN SPACE PROGRAMME

After pursuing his childhood interest in rockets and engineering, Jean-Patrice Keka started to formalise his research into a single programme – Troposphere – the Congo's first aerospace company.

JEAN-PATRICE KEKA: After university, I thought, *I now have the knowledge that will enable me to do something tangible*. Other countries had started similar programmes and were already reaping the economic benefits, so why couldn't we do the same? This led to me formalising my ideas in a document I called 'Aerospace Program for the DRC'. I also named the programme 'Troposphere' after the atmospheric layer that is the closest to the Earth's surface. I was still at the research stage, and the rockets I'd made on my own couldn't have gone higher than that. During that time, there was no official Congolese space programme, so my booklet describing the Troposphere programme was really the beginning of one.

Even when I read my document now, I find it clear – the ideas were good – and at the time I thought, *I'll follow this*. But I didn't know where I could [put my ideas into practice]. That's when I realised there were rockets I could make without a proper workshop. I lived in a flat in an area of Kinshasa called Lingwala, and there was some room on the ground floor, so I took up that space.

People could see I was building something, but I didn't want them to see it was a rocket. If they'd found out, it would've been a real bother, and I wouldn't have had time to focus on my task. More importantly, if anyone from intelligence or security had discovered I was building a rocket at home, they'd have immediately thought about explosions, and I'd have been arrested, because I might have been jeopardising people's lives.

Since I didn't want people to know what I was doing, I decided to build only the body on the ground floor, and I built the rocket's tip in my flat. Then, when people were away, I'd hurry downstairs to test the tip with the body before quickly returning back upstairs with it. I didn't assemble all of the parts until the day I decided to transport it. During that time, a government minister who was also a physicist came to visit me and said, 'What are you doing here?' I made up something else, but I was actually building Troposphere 1.

I worked and worked on Troposhere 1, but it was made from PVC. I didn't want [the material] to fail, so, once it was finished, I moved onto Troposphere 2 which I built out of powdered milk cans. I dream, but I'm also realistic, so I use what I have and make what I dream out of what I've found. That's my core principle.

Once I had my two rockets in hand, I had a little money, so I found a location in Menkao, and I called the authorities who could authorise it to ask for permission to launch. The authorities didn't even understand what I was saying: 'What do you mean a rocket? Stop with this business.' I thought, *I'm not going to find a solution here – I just need to go.*

Journalists would come by sometimes, but I couldn't tell them what I was doing, because they might have broadcast it, and I would have got into trouble. But now I told them, 'They're rockets. I'm going to Menkao to launch them.' Many of them made arrangements to come with me, and I thought, *I'm somewhat protected. If I'm arrested on the way, it'll be in front of cameras.*

The next problem was transportation, because if you tell someone you're putting a rocket in their vehicle, they might think it is going to explode

Right: A liquid propellant tank in a supersonic ramjet engine undergoing testing for Troposphere 3.

Below: The avionics of the Troposphere 4 rocket.

and damage it. So, I had to conceal it and say it was something else and not a rocket. I rented a lorry that belonged to my landlord, and I put the rockets' bodies in so you could only see the tips and pretended I was taking something to my farm. Thankfully, he believed me.

I was also scared that the lorry carrying the rockets might be stopped and that people would think I was launching or testing missiles. On the road, there were small roadblocks to check what you were transporting, so I'd told everyone who was coming that if we ran into the police, they had to stay silent and let me do the talking.

When we got to the launch site, a lot of onlookers came to see what was going on. I also told the village chief what was happening, because even though I'd bought a plot of land, back home you always needed the village chief's approval. When I went to inform him that I was testing a rocket, he had never seen one before. No one had.

A launch day is always a marvellous day, wherever you are. In the USA,

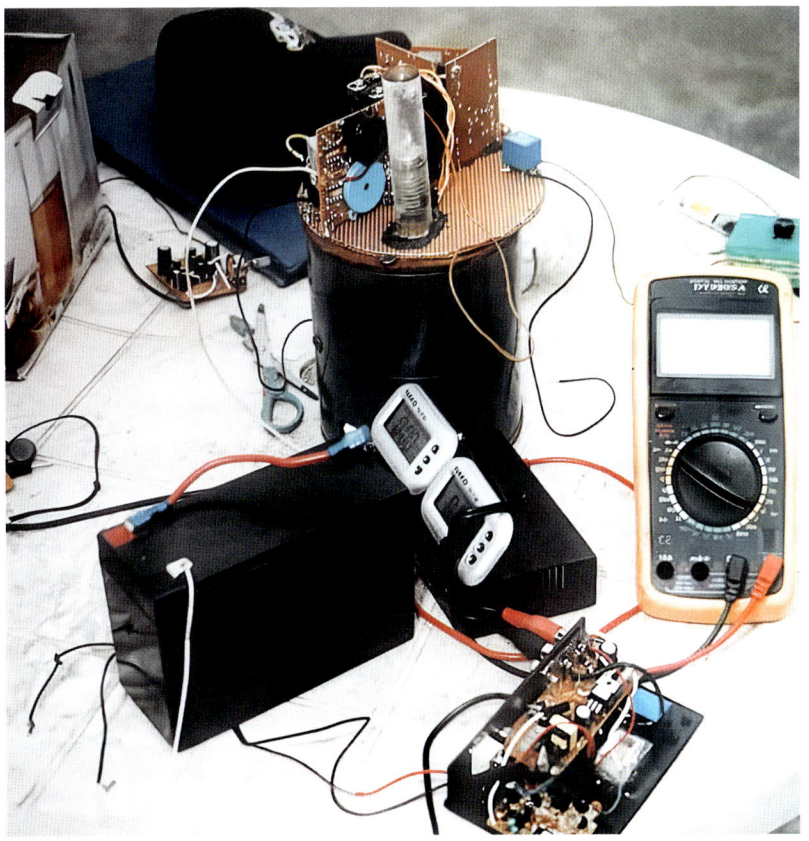

you can even pay to be closer to the action. So, it was a momentous day, because I was setting up a proper rocket for the first time.

Some fellow scientists were there to watch, and you could see on their faces that they didn't know what to expect. I'm a Christian, so I knew God had given me this talent, and it was going to work. I was very sure. But the first rocket had an issue. The monitors were about 100 to 200 metres away from the launchpad in a sort of closed hangar, and I did the countdown myself there in front of all the journalists and their cameras. When I got to zero, I pushed the button and saw smoke, but Troposphere 1 didn't take off. It was raining, and it wasn't protected well, so I realised immediately what the problem was: the fuel had sucked up some raindrops. It was like petrol in a car – it doesn't work if you add water.

When the first rocket failed, the journalists laughed and made fun of me. But that's how science goes. They were asking, since the village chief was here, 'Did you prepare a drink for the chief or our ancestors so they could give you the green light?' I told them that had nothing to do with it. Here it was all about equations.

I then asked some friends who'd come to quickly set up the second rocket, after which I would go myself and make sure everything was connected correctly. I was just on my way when the rocket took off. All the journalists fell over, because they were scared seeing this heavy thing in the air, and some weren't even able to capture any footage. In fact, almost everyone fell over except me, because I knew what was happening. That day was an extraordinary success. The spectators hoisted me in the air, carrying me like a little kid. It was an overwhelming feeling of joy. I don't know how else to explain it.

After an achievement, we always forget about equations and dance like madmen and transform, just as we've transformed the planet. Then we go back to reality and our equations the next day. I danced a lot that day!

It was a very significant day for me, because it signalled things beginning in an official way. And it impacted my entire life, because I saw that I could make it work. It also struck me as a big moment for the country. These were the first steps that Congo was taking in the aerospace sector.

Some people see me as a genius; some see me as a madman. Others call me a dreamer, albeit a dreamer who makes things that work. I accept it all. My life really is a great adventure. And it's a contagious kind of adventure. I now have 50 mathematicians, physicists, chemists and engineers who believe in this adventure too. The first thing I'm looking for in someone is for them to come willingly because they love this. If they say, 'It's my passion,' that means a lot to me, because I know that by my side they will change and learn. That's how I work. You have to love it. And together, we're trying to open closed doors and striving to cross frontiers. My volunteers are extremely important to me, and we've become like a family.

Right, and overleaf: Photos from Jean-Patrice Keka's Troposphere 4 launch, attended by the government minister for science and village chief, 2008.

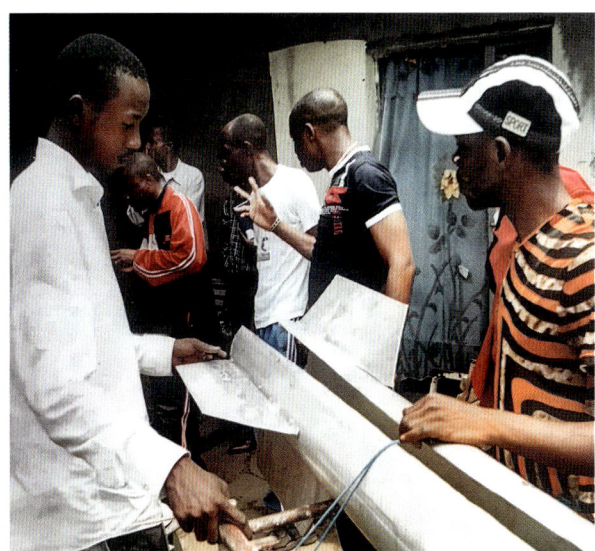
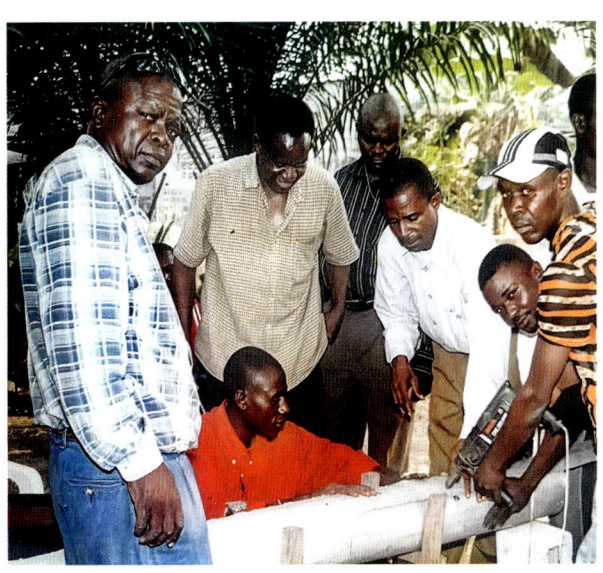

After Troposphere 3 failed to launch, Keka turned his attention to the launch of Troposphere 4 in 2008.

JEAN-PATRICE KEKA: For Troposphere 4, I called the government minister for science at the time. He understood the importance [of what I was doing]. I'd met him before, and he'd helped me, so I invited him to the launch and said, 'Do us the honour of pressing that little button so we can see if the rocket takes off.' He brought along his enthusiasm and his entourage. It was the first time I was hosting such a big figure, and it really was a great honour and joy for everyone. When he pressed the button after the countdown and the rocket took off, he was extremely surprised that it had worked.

On the psychological side, it was extremely important to see things we'd picked up and assembled starting to bear fruit. The minister accepting our invitation despite his rank, and coming to see what we were doing, boosted us and gave everyone strength and the belief that we were on the right path; it meant the authorities were starting to understand us.

My philosophy is to go further after each launch. That is, the second rocket can't be like the first one, and the third rocket can't be like the second one. This means increased engine power, increased altitude, increased weight, and so on. That's how I've progressed. If it was all proportional, we'd have multiplied Troposphere 4 by four, but sometimes we increase it even more. [With Troposphere 4,] we reached a 15-kilometre altitude with a rocket weighing more than 250 kilograms, and we achieved a velocity of more than twice the speed of sound. That's pretty good.

Watching a rocket being successfully launched is the most emotion you can feel. It can even overwhelm you. You can't help but think, *Is it going to work or not?* And there's the safety part too. You could kill people and yourself. All those elements mean that pressing the button is a great responsibility. When we invited the minister, we were asking him to press a button that might create an explosion and kill him.

The minister was the first [powerful person] to share our vision. He said, 'To all the other African countries, come and join us so we can take the continent to space!' So, it started with the minister beginning to believe, and he probably spread that belief to a lot of people. [Troposphere 4] helped people gradually to see that we could succeed and that it was possible. I felt very joyful.

At the start, in my youth, I wasn't thinking of the country. It was a passion first, and I wanted to fulfil that passion. That's it. I didn't think about finding solutions for my country, because I didn't know what solutions were needed, since I was too young. I started to have ambitions for the country when I developed more academic knowledge, and I began to grasp the importance of rockets and what they could bring.

People told me I should create a framework in which I could give classes and explain what I was doing to people, so my knowledge didn't all disappear with me one day. I already go into schools whenever I get the opportunity and explain about rockets and such. I ask the children, 'Do you want to be like me and do what I do?' It's that simple. They go home and say, 'Mum, Dad! I want to be like Keka!' 'What does this Keka do?' 'He launches rockets!' 'You want to launch them too?' 'Yes, I want to be like him!'

The goal is to inspire those children in schools so a younger generation can continue my work when I've gone. But not just as replacements. If they make fast progress, they can become our colleagues. They might even become more efficient than us. The country needs it.

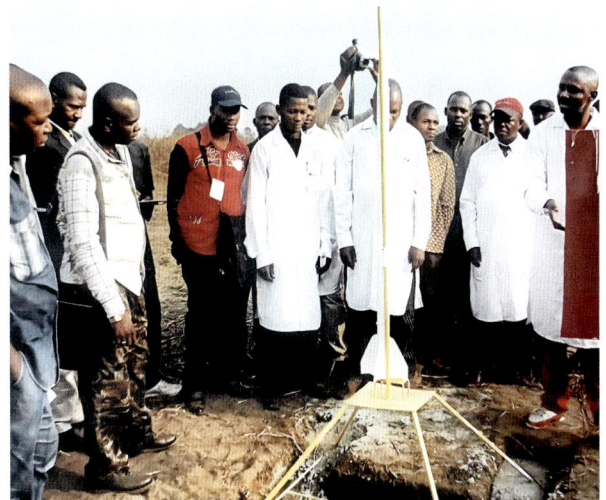

Troposphere 5 was launched a year later, in 2009.

JEAN-PATRICE KEKA: Tropospheres 1 to 4 were single-level rockets, but with Troposphere 5, we wanted to produce a two-stage rocket, which would allow us to reach our 36-kilometre goal with a cluster of several small engines that would provide more thrust. But that technology is very, very complicated. A lot of great countries have failed to manage it. We therefore made every effort with Troposphere 5, because we knew it'd be very challenging for us, but we had to try it.

Unfortunately, the rocket fuel wasn't the right consistency, and when we transported it, because it was a solid fuel, it took a different shape. Then, when we got to the launch spot, we couldn't get the igniter to go into one of the solid-fuel canisterss because the fuel wasn't aligned properly anymore. In the end, one of the canisters exploded, changing the rocket's trajectory so that it landed 500 metres away. Fortunately, it didn't hurt anyone.

When I later said the launch was a success, people laughed and didn't understand. But it was a great success for me, because all the other canisters held on! Even the canister that exploded only did so after a struggle. So our steel choices, our thickness calculations, all this and more were right. And the rocket even took off! So, to me, it was a success.

The Troposphere 5 launch became world news, and when we posted the video on the internet, we reached a million views in a heartbeat, because it was the first time people had heard of a proper rocket in Africa. Also, it wasn't a small rocket anymore – it was a two-stage rocket that weighed almost a ton. So, a rocket of that weight, with the basic boosters having to lift all of it and take it beyond the speed of sound, was getting really serious. The experts could see this, but some non-experts might have thought, *They're Africans. They can't do this. They need to let it go and move on.* So, there were different types of interpretations. Some people were encouraging, others were mocking us, and we had to accept it all.

[Some of the negative comments included:] 'A horizontal trajectory, but you call it a complete success! Epic.' 'A Congolese engineer? That's an oxymoron.' 'Finishing touches done with an angle grinder . . . They aren't professionals.' 'Just a little more effort, guys! In no time, you'll have drinking water at home, or even flushing toilets.' Some of them assumed 'people from Africa can't create anything worthwhile'. But this is ignorance, as it has nothing to do with skin colour. It's just science, and science has no colour and knows no borders.

Everything I read can be summed up as people not understanding how difficult it is to do what we do. For those people, all that counts is complete success. But whether the rocket is a 100 per cent success, a partial success or a failure, it'll give us a lot of information. But I'm immune to the ridicule. I know what I'm doing. I know where I'm going. So it didn't shake me. On the contrary, it strengthened me.

I put a lot of care into the work, and Troposphere 5 didn't fail because I wasn't capable. Simply put, there were a lot of moving parts. But I have self-confidence, and I've never lost it. That's why I still believe that Troposphere 6 will go up and do what I want. And I believe beyond Troposphere 6 that we can make our ideas happen. And I'm now surrounded by people who get it, who share my philosophy. I don't say 'I' anymore; I say, 'We'll all achieve what we believe in.'

After the launch of Troposphere 5, a lot of big space agencies contacted us, and we're still working with some. We want to get even better, and they're going to help us progress further. So, you could say that experts are gaining interest in our work. When we started launching rockets, we were getting attention from within our country, but now it's really become international.

The Indian space programme started small too and endured difficulties, but it has reached a very high level. When our Indian friends were beginning, they'd transport their rocket parts on bicycles. It's exactly like back home. It's very close to what we're doing. And we think, despite our difficulties, that one day we'll also reach as high a level as them. We will succeed. People might've laughed at them and called them naive, but they're not naive anymore. We are destined to get there too, and we've already taken a step, because Troposphere 6 is going to be bigger.

Right: The launch of the Nike Apache rocket, the first rocket launched by India from the Thumba Equatorial Rocket Launching Station in Kerala, 1963.

THE NEW SPACE RACE

Top left: DRC rocket scientist Jean Patrice-Keka shows off his drawing of the Mpongo spacecraft, designed to transport a galaxionaut and release at an altitude of 45km.

Top middle: An industrial design.

Top bottom: The second stage of Troposphere 6 during the construction process.

Top right: Jean-Patrice Keka describing his Troposphere project to the media.

Top middle and bottom: Jean-Patrice Keka with his prototype Troposphere rockets.

THE FINAL ASSEMBLY OF THE ISS

While the Columbia disaster in 2003 caused setbacks for the assembly of the ISS, flights resumed in 2005. Terry Virts was the shuttle pilot on mission STS-130, which flew to the ISS on 8 February 2010 to install the station's Cupola and Tranquility module.

TERRY VIRTS: On my first space flight on the Space Shuttle Endeavour, we finished building the ISS. We brought up two modules: node three, which is a big living module [also know as the Tranquility module], and the Cupola, which is everybody's favourite place in space. It's a really cool module with six windows around you and a seventh big window above you. It's on the bottom of the space station, so when you look up, you see Earth.

I love the *New York Times*'s description: 'a new window that forever changed our view of Earth'. It changed my view of Earth, for sure. Because I assembled it, I was the first one to open the window cover of the Cupola. The other windows are hatches that you look out of, but you put your whole body inside the Cupola, so you're surrounded by the galaxy. It's really an amazing experience. It makes you feel like you're on a space walk. As I was in the Cupola looking out, I had this emotional feeling: *That's my planet over there, and I'm not on it. I'm in space.* It was a really profound moment for me.

Below: NASA astronaut Terry Virts takes Earth observation photos from the Space Station's Cupola in the Tranquility module, which his Shuttle mission STS-130 has just delivered to the station. Taken in 2010.

THE END OF THE SPACE SHUTTLE

The Columbia tragedy was a key factor that led to President George W. Bush announcing the end of the shuttle programme in 2004. After work on the ISS was finished, Space Shuttle Atlantis completed its 33rd and final flight on 8 July 2011. Kristin Fisher, daughter of astronaut Anna Fisher, was now working as a space journalist.

For the first time since the 1981, America lacked any means of sending their astronauts to space.

KRISTIN FISHER: I was there covering it for work, but I got to watch the actual launch side by side with my mom. There was a camera filming us, and I was kind of subdued. I was trying to keep my reaction in check, but, true to form, my mom was cheering, hooting, hollering – classic Anna Fisher. She was also crying, and I was pretty tearful too, because despite Challenger, despite Columbia, despite all the tragedy that the space shuttle had been through, it had also been an unbelievable programme and an incredible vehicle.

A lot of people were really worried and fearful about these gap years in which the USA was going to be reliant upon Russian Soyuz rockets to take their astronauts to the ISS – for a lot of astronauts, it was a really tough pill to swallow, and there was deep concern and fear about where NASA and human space flight in the United States was going to go.

[At the time,] my mom was working on a very early version of the Orion spacecraft. Back then it was called the Orion Multi-Purpose Crew Vehicle. Now it's just Orion. This is the vehicle that is going to take astronauts back to the Moon some day. But it was so far away from being done, and I think everybody knew there was a real lack of political will in the USA around that time to fund the human space flight programme in the way that it needed to be. I think that's why there was so much sadness around that last shuttle launch. There was no sense of urgency that its replacement would be ready.

I still think it was the right time to end the shuttle programme. Those vehicles had proven to be deadly on more than one occasion, and there was a widespread acknowledgement that it had reached the end of its lifespan. There were also safer options out there. As much as the pilots and my mom loved the fact that this was a winged spacecraft that was reusable and that you really got to fly, all those things made it a lot more dangerous too.

Above: A spectator shields himself from the sun with a commemorative fan at the final launch of the Space Shuttle Atlantis at the Kennedy Space Center in Florida, 8 July 2011.

TERRY VIRTS: When the shuttle retired, I was sad. I mean, I loved flying it. I wanted to be a shuttle commander. I wanted to fly more space shuttle missions as a pilot. The height of my career was being a space shuttle pilot. But the reality was that we had killed seven of my friends in the Columbia accident, and the shuttle was something that could only go into low-Earth orbit, whereas I thought we should be going back to the Moon and Mars and doing deep space exploration.

Becoming completely dependent on the Russian Soyuz felt like a really bad idea. At NASA, we always want back-ups. We'd had a space shuttle accident that had prevented us from launching anybody for years. So, what if the Russians had an accident? It had been a long time, but they'd had a lot of close calls, so if that had happened, we would have had no way to send astronauts to the space station.

Right above: Space Shuttle Endeavour makes its final flight atop NASA's Shuttle Carrier Aircraft as it comes in to land at Los Angeles International Airport on 21 September 2012; on its way to be displayed for visitors at the California Science Center.

Right: President Obama inspects the mission patch, during a meeting with the final shuttle crew.

Overleaf: The final Space Shuttle mission, Atlantis STS-135, clears the tower on 8 July 2011.

THE NEW SPACE RACE

POLITICS ON THE ISS

Below: Expedition 35 Russian Flight Engineer Alexander Misurkin is helped into his Russian Sokol suit before his launch on the Soyuz rocket to the ISS at the Baikonur Cosmodrome, Kazakhstan, 28 March 2013.

In 2013, Russian cosmonaut Alexander Misurkin travelled to the ISS as a crew member of Expeditions 35 and 36.

ALEXANDER MISURKIN: My desire to become a cosmonaut came from my childhood dreams about space. They were based on watching the *Star Wars* series, which made a huge impression on me as a boy. In the early 1990s, [when I had to think about my career after school,] I remember asking myself, *What is it that I'd like to do?* The image of a Jedi came into my mind, but I quickly realised this wasn't an option. I was disappointed to realise that I'd never fly the Millennium Falcon. Yet flying to the near-Earth orbit was more of a realistic prospect, and so I thought, *I want to be a cosmonaut.*

Firstly, I wanted to go where no one else had been before. I knew that a number of people had already been in the near-Earth orbit before me, so my hope was, perhaps I could fly to the Moon, or to Mars. Secondly, I felt a need for personal development, so I thought this profession would give me the opportunity to do that, and I was not wrong there. Thirdly, I heard there was an international project underway, so there could be a real opportunity to see what life in other countries was like, with my own eyes.

The ISS began to be permanently inhabited from the year 2000. I was selected into the corps in 2006, and began my training when the ISS project was already up and running. My first mission, Expedition 35, took place in

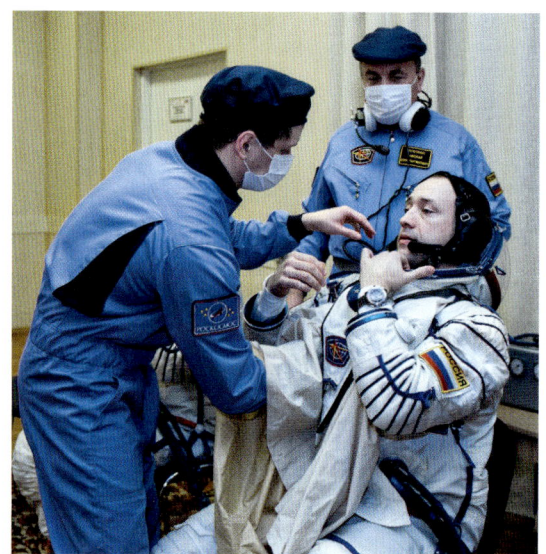

2013, meaning there had already been 34 expeditions onboard the ISS. I didn't see myself as a pioneer or initiator of this project. I felt as though it had been tried and tested, so it was more of a conveyor belt, and I would come on board as the latest link in the chain. I felt that by my nature, I am someone who likes to live in a harmonious, peaceful and orderly society, so I thought we had to specifically apply this way of living to our relations with our crew members. It was only when I was on my second mission, this time as expedition commander, that I fully recognised that this was my responsibility: I needed to organise our life on board in a way that promoted respect and friendly relations.

For the most part, it makes no difference whether the team is international or not. Conditions during any space mission are life-threatening at all times, you are constantly aware that you're in a hostile environment. This serves as a seriously unifying factor for the entire crew.

Contemplating our Earth and how fragile it is, adds some extra gravity to this need for unity. I always understood my life on board the ISS as being part of a tiny united society. This is probably why I have the view about the world that I have, that our life on Earth should also be like this: a single society, living in harmony.

Below: A pre-set electronic still camera (ESC) captured this traditional in-flight crew portraits for the STS-88 members on Endeavour's mid deck. From left to right: Frederick Sturckow, Jerry Ross, James Newman, Nancy Currie, Robert Cabana and Sergei K. Krikalev. A banner representing the participating countries for ISS and a model (near Krikalev) of the connected Unity-Zarya modules are in the background. The STS-88 carried the first US part of the ISS to orbit (the Unity Node).

GINGER KERRICK: Politics can play a factor in our day-to-day lives here on Earth, but we did not want it to play a factor in operations on board the ISS. For example, we had conversations with our crew members about our expectations if a war were to break out between countries that corresponded with the citizenship of the astronauts on board. While there are certain things we can prepare our crews for, we can't predict what one nation is going to do against another nation on the ground. The only thing we can do is equip our crews with the right mindset.

THE NEW SPACE RACE 287

In 2014, astronaut Terry Virts began a long-duration mission on the ISS, as part of Expedition 42. It was during that six month mission that he witnessed the escalating political discord between his and his colleague's nations on Earth.

TERRY VIRTS: We all had our heads buried in the sand, and we were ignoring the really profoundly bad things happening in Russian society with Vladimir Putin. My concerns at the time were about safety and not having a replacement for the shuttle – more so than the geopolitical situation. But we should have seen that coming, and we didn't. I think everybody at NASA was so excited about building the space station together, about cooperating with our former enemies, that we were willing to ignore the warning signs.

In the months leading up to my launch, I was training in Russia when Russia invaded Crimea. The geopolitical situation was tense, but Obama didn't react. There was no American pushback at all. So, even though it was tense, it really wasn't a big international conflict yet, because America just let it happen, and we didn't really worry about it. As we went through training, at the end of each course, there'd be a party – the Russians are always partying – and we'd get the vodka out and make a toast. And the first toast would always be along the lines of, 'Politics is politics. Here's to the crew.' The sentiment was that we were going to try to stay alive in space and not worry about all these political things on Earth. I was blinded by the excitement of having these Russian partners, and I did not personally understand the magnitude of the situation.

Then, in March 2014, just a few months before my launch, the USA put sanctions on Russia because of their annexation of Crimea. Until that point, I had been pretty unaware of the severity of the situation, but I started to think maybe there was something going on here that wasn't good. When we sanctioned Dmitry Rogozin, who was the head of Roscosmos, the Russian Space Agency, he got mad and tweeted something along the lines of, 'Hey, Americans, you can take a trampoline to get to the space station.' Our only way to get to space was on a Soyuz, so he was threatening to take away our ability to get to space. When he tweeted that, I thought, *We're buying rocket engines from Russia. We're doing all this cooperation. This is really going to hurt Russia in the long run.* And that's exactly what ended up happening. NASA immediately issued contracts to start building American spacecraft and rockets so they would no longer be dependent on Russian launches. We also accelerated the commercial crew programme so we could launch astronauts on our own rockets and not on the Soyuz.

Left, top: Terry Virts in his Russian Sokol pressure suit.

Left, bottom: A meeting of the minds aboard the ISS on 7 March 2015 with members of Expedition 42: at the top upside down is astronaut Barry Wilmore (mission commander), to the right is cosmonaut Elena Serova, and ESA astronaut Samantha Cristoforetti. Bottom center is US astronaut Terry Virts, next to top left cosmonauts Alexander Samokutyaev and Anton Shkaplerov.

Below: Photo of two Soyuz spacecraft docked to the ISS, sent via Twitter by astronaut Terry Virts during a 6-month expedition on the ISS, 22 February 2015.

GINGER KERRICK: When I saw the news that Russia was annexing Crimea, I thought, *Is this going to destroy what we created with the space station? Is somebody going to come in and tell us that because the country we've been partnering with peacefully for so many years has declared war on another country we can no longer work together?*

When you talk to people who were inside of mission control, or when you talk to the crew on board the station at that time, they say it was something that was happening outside. There was a protective bubble around them, and they were unified by their mission. They realised that it wouldn't be a punishment to the other country to pull out. It would be a punishment to both. There would be no winner in that scenario, because neither country could operate that spacecraft without the other. No matter what was going on outside, they had to keep the crew safe, maintain the integrity of the spacecraft and execute the mission. Back then in 2015, outside we faced challenges, but inside that bubble it was OK.

TERRY VIRTS: In January 2015, I was in the Russian segment of the ISS, in the service module, with Sasha Samokutyaev. We were sitting there looking out the window when we flew over Ukraine, and, all of a sudden, I saw these little red flashes on the ground. We were watching Ukrainians being killed by Russians. It was a profound moment. I thought, *Holy cow, I can't believe what I'm seeing*. It was probably the most poignant moment of my time in space. When Sasha and I saw those bombs go off, I thought we both had the same reaction: *This is terrible. War is bad. This needs to end.* But we didn't talk about it. I just assumed that was the conclusion he and the other cosmonauts would come to. I was shocked to be actually watching war from space. But the absolute, overarching culture of the space station was to keep politics out.

It was an incredibly strange time of cognitive dissonance: on the one hand, *I love these guys. I'm training with them, I'm excited to fly with them*. On the other hand, *Their government is doing this terrible stuff in Ukraine, and Chechnya, and Georgia, and Transnistria, and Syria*. It was a learning experience for me. I had just not put everything together when the invasion first began in 2014.

ALEXANDER MISURKIN: We had a rule, if you're going to Houston, there were three things you don't talk about: politics, religion and money. Until recently I had no interest in politics at all. And I simply had no interest in talking to other astronauts about it. Yet, I know for a fact that I have friends — cosmonauts and astronauts — who did discuss these things quite openly, because they had broader interests, and they found learning about each other's views interesting.

So I'd say this is a matter that's decided on an individual basis. I believe that key to resolving any conflict is staying respectful towards each other. If you are talking about subjects where you may have differing views, but you stay respectful, there shouldn't be any problems discussing them.

TERRY VIRTS: That fun, optimistic excitement that we had in the 2000s changed to a more subdued reality. When Russia did the full-scale invasion in 2022, it was obvious that the space station, which I loved and was something I risked my life to build and operate, was not nearly as important as what was happening in Ukraine. The age of innocence was gone.

Below: Group selfie of participants of Expeditions 42 and 43, including Sasha Samokutyaev and Terry Virts.

SERGEI ZALYOTIN: Fundamentally, [personal feelings about the war in Ukraine] haven't affected Russian cosmonauts. They were flying and they're still flying. If it was affecting these cosmonauts, they should be removed from the squad and new ones should be recruited who wouldn't be affected. I don't understand how anyone can let political decisions beyond their control, impact their professional activities. If [it does] then the question is, 'Do you have what it takes to be a cosmonaut?' If any American astronaut is affected by the political relations between states, I'd be fascinated to hear what they have to say but in my opinion that astronaut should retire.

I wasn't the one who started this war, and the cosmonauts weren't the ones who started it either. We each have our own point of view but we cannot let that impact the flight programme or safety of the crew onboard the ISS. When the Americans bombed Libya or Yugoslavia, it had a huge impact on me, and I was agitated, but no one ever asked me about that. My personal concern didn't matter.

ALEXANDER MISURKIN: We were affected by [the Russia–Ukraine war], and it took time for me to determine my own position as to what was going on. It is clear that any war is a terrible thing. It is evidence of the fact that the two opposing sides have been unable to find not just a mutually beneficial solution, but any solution that might have respected the interests of one side or the other. For me, like for the majority in Russia, it was a shock that this happened. And I think that [shock was felt] not only in Russia but throughout the world too.

Before February 2022, there were no major tensions [on board the ISS] in the relations between our nation states, and it was simpler to live as a united crew. After 2022, I understand that there are crews that are more disparate in their day-to-day lives, and there are crews who, despite everything, continue a more amicable life on the ISS and who are more open with one another. Of course, I side with the second approach. I am convinced that national rifts, conservatism and throwbacks to divisions in our global civilisation are all backwards steps.

As tensions between Russia and the US continued to escalate over the invasion of Ukraine, in mid-March 2022 a tongue-in-cheek video was posted on social media by Russian government-controlled RIA Novosti. The video showed NASA astronaut Mark Vande Hei being left behind on the ISS by his cosmonaut colleagues. Vande Hei was sent up to the ISS on a Russian Soyuz capsule in October 2021 and was due to return to Earth on the Soyuz on 30 March 2022.

The video garnered international attention when the then head of the Russian Space Agency, Dmitry Rogozin, posted it on X. Since its publication, the video has sparked varying reactions from the US and Russia as to how serious the message in the video was.

TERRY VIRTS: After the war started, [former head of the Russian space agency] Dmitry Rogozin published a video of the space station splitting apart and leaving [US astronaut] Mark Vande Hei stranded in space. Your partner threatening to leave your astronaut stranded in space was immoral – it was ridiculous. It started to click in my brain that there were geopolitics on Earth that had not affected the space station before that were now going to have an impact on space. Rogozin had corrupted the space Garden of Eden.

In the wake of Russia's invasion of Ukraine, then head of the Russian Space agency, Dmitry Rogozin, responded to sanctions by threatening to de-orbit the ISS via a series of tweets online.[2]

SERGEI ZALYOTIN: For me, [Rogozin's threat] was quite foolish and more of a political show than a real attitude towards the work we were doing on the ISS. You see, to de-orbit the ISS is a very complicated procedure, so this was not a genuine threat. We were all mad at Rogozin, and I was mad at him, but what can we do?

KRISTIN FISHER: When you work for a company like CNN, you have access to a truly global news-gathering organisation. So, when I first started there, I thought, *If I could interview anybody in the global space industry right now, who would it be?* The answer was immediately Dmitry Rogozin.

I reached out to some contacts at NASA to see if they thought he would play ball, and the response I got was, 'You know, he might. He's kind of a loose cannon.' I was shocked that he agreed to do it, but it was a fascinating interview. Basically what he said was, 'We don't think it's fair for the USA to impose sanctions on the very entity that it's also a partner with up at the ISS.' But when I pressed him on whether he was serious about his threats, his exact words were, 'Divorce is not possible at the International Space Station.'

It was a Zoom interview, so I never got to sit down with him in person, but my big takeaway was that I didn't realise how much of a jokester he was. He had a lot of pretty good one-liners in Russian for the US audience. My other big takeaway was that a lot of what he was saying at the time was just bluster, like threatening to leave NASA astronauts behind on the ISS.

The ISS is remarkable in the sense that it has been a beacon of diplomacy for more than two decades, and it has survived multiple crises on Earth. But the invasion of Ukraine stressed the partnership more than it ever had been before. At one point, I remember not knowing of any Americans in Moscow, except for the people who were in Star City, working in Russian Mission Control. The fact that they were able to continue doing their jobs, and the space station continued to function throughout the conflict, is pretty remarkable.

---

[2] On 24 February 2025, Rogozin took to Twitter, warning NASA and the US: "If you block cooperation with us, then who is going to save the ISS from an uncontrolled descent from orbit and then falling onto the territory of the United States or Europe?" he wrote. "The ISS doesn't fly over Russia, so all the risks are yours."

Above: Cosmonauts unfurl a Soviet-era victory banner on a recent space walk. The banner, which is used to mark 'Victory Day' in Russia, has also been used frequently by Russian forces in Ukraine.

THE NEW SPACE RACE

THE RISE OF SPACEX

After the retirement of the space shuttle, NASA was reliant on Soyuz rockets for transporting their astronauts to the ISS. To sever this dependency on Russia, NASA turned to the private sector to help develop an alternative launch vehicle.

GINGER KERRICK: [The end of the shuttle] caused us to turn our focus to how we could help the commercial space industry. NASA put out a bid, and seven companies responded. We were trying to help all seven get to the next phase, but in 2014 NASA selected SpaceX and Boeing and put all their eggs in those two baskets. Both companies were supposed to fly an uncrewed and then a crewed test flight to the space station. In 2019, SpaceX sent their uncrewed test flight, and it was awesome. No problems at all. They were then scheduled to launch [their crewed test-flight in] May 2020, but in March 2020 COVID hit and everything stopped.

I couldn't send my crew to Moscow to train, so I called my Russian counterpart and said, 'What are we going to do? I want to hit this May launch date. We can put our crew members on a NASA plane and get them to you. What can you do?' They figured it all out, and we were able to launch that test flight crew on time in May 2020 and bring them back safely.

By the time SpaceX was flying its first crewed test flight, I was no longer a CapCom and no longer in the flight director office. Instead, I was managing a division that was responsible for the operational safety of our crews, so I was very engaged, as this was a crewed test flight. I wanted to be able to look Bob Behnken and Doug Hurley straight in the eyes and say that I had faith that this vehicle was as safe as human space flight gets.[3]

We had been talking about launching a US crew from US soil for so many years that after a while, you start to convince yourself that it is never going to happen, that we've lost our lead in the human space flight industry, and we aren't going to have a crewed vehicle. So, there was a sense of, *Thank God. Now we have a vehicle that's capable of sending crews.* It made America a spacefaring nation again. It was the beginning of a new era.

---

[3] NASA astronauts Bob Behnken and Doug Hurley flew on Crew Dragon Demo-2 on 30 May 2020, the first crewed SpaceX flight to the ISS.

Left, above: Crew Dragon Demo-1 mission patch.

Left, below: SpaceX Crew Dragon Demo-1 sitting inside the LC-39A Horizontal Integration Facility in Florida, next to a Falcon 9 rocket.

Below: Launch of the first test-flight of SpaceX's Dragon Crew spacecraft, Crew Demo-1, to the ISS on 2 March 2019. This mission did not carry humans; the first crew launched on Crew Demo-2.

THE NEW SPACE RACE

KRISTIN FISHER: SpaceX is so successful because they're not afraid to fail. They see that pushing their spacecraft to the limits is a good thing in the testing process – they call it rapid iterative development, in which explosions and failure are normal. But another reason they've been so successful is that they're not afraid to completely redesign everything we know about rocketry from the ground up. The Space Launch System, on the other hand, is a Frankenstein rocket.[4] It is literally cobbled together from space shuttle parts, including their main engines. SpaceX has rethought everything, from the fuel to what the rocket is made of to how they streamline the design of the engine. That's why they were able to develop reusable rockets long before anybody thought that was scientifically feasible.

SpaceX is now the number-one launch provider in the world. There is no company or country that can compete with them in terms of the speed and reliability with which they launch people and payloads into space. Think about that. Not the USA, not China, not Russia, can compete with what this private company is doing, and Elon Musk sits on top of it all. It makes him more powerful than anybody can fully comprehend at this moment in time. Then, on top of that, you have Starlink, the mega constellation of small satellites that SpaceX is launching into space on average once every three days, providing global connectivity, eventually from your cell phone. People who follow space have seen the speed with which Elon Musk and SpaceX have been able to amass this global dominance. The US government hasn't been able to keep up. He has even created a super heavy lift version called Starship.

Launching a spacecraft and the booster then returning to Earth was stuff that you read about in science-fiction books, but Elon Musk made it a reality. He was ridiculed for the idea that you could have a giant launch tower with big mechanical arms – SpaceX call it Mechazilla – and that this massive, super-heavy booster would then return to the very launch pad that it had just lifted off from and these arms would catch it in the middle, like chopsticks. They did it on their first attempt.[5]

I woke my daughter up early to watch the Starship launch with me, and I was trying to explain the significance of the moment. The countdown started, and I was cheering when my daughter said, 'Mom, it's early. Stop cheering. What are you doing?' I thought, *Oh my God, I've become my mother*. Then when it actually caught the super-heavy booster, I lost it. It was the coolest thing I'd ever seen. I go back and watch it now, and I still can't believe that it happened. But I completely turned into my mother in that moment – it was a full-circle moment.

---

[4] NASA's Space Launch System is designed to launch the Orion spacecraft on a trajectory to the Moon, and as of 2025 has had one successful uncrewed launch.

[5] SpaceX successfully caught their Starship Super Heavy rocket booster as it returned to the launch pad on the rocket's fifth test flight on 13 October 2024.

Below: SpaceX Starlink Falcon 9 rocket launch as seen from the National Geographic Venture in the Sea of Cortez, Mexico.

THE END OF THE ISS

The ISS is scheduled to be decommissioned by the end of 2030. After initially requesting a withdrawal from the project in 2024, Russia has agreed to continue its commitment to the station until at least 2028.

GINGER KERRICK: Studies to extend the life of the space station [are being carried out], because we want to make sure that we don't have a gap in low-Earth orbit [for human habitation], and NASA has put forward some seed money towards commercial space stations. The goal is: do not retire the International Space Station before you have a viable alternative up in space. But some of those plans are experiencing delays, so I think NASA intends to keep the ISS up as long as it is safe and viable to do so, to either have no gap or at least minimise the gap with the private space stations.

It would be a tragedy not to take advantage of what we've built and take it to the next level. Logically, it makes no sense not to continue. And I like to think that sometimes logic prevails, that someone will look at the plans that we have moving forward and recognise the value that collaboration brings. It's not like we haven't proven it. So, once the dust settles on all the animosity, maybe people will circle back and say, 'There is value here. Let's continue to collaborate, capitalising on each other's strengths.'

KRISTIN FISHER: The end of the ISS is imminent, but not necessarily for political reasons. Yes, the relationship is clearly wearing thin, and has been for some time, but the real impetus for the end of the ISS by the 2030 deadline is the sense that this is an old orbiting outpost, and it's starting to show signs of real wear and tear. There have been multiple leaks, and astronauts and cosmonauts have been doing more and more repair missions. So, the reality is that it's reached the end of its lifespan. And most people think that the partnership between the Russians and the USA will likely come to an end when the ISS is deorbited.

ALEXANDER MISURKIN: As I understand it, [the end of the ISS] is primarily linked to the differing views on how much longer the Russian and American segments could be used. I would like to believe this is more of a technical matter than a political one. The ISS was launched in 1998, when its initial lifespan was set at 15 years, at least in respect of the first modules that were made in Russia using American money. As of today, its lifespan has been almost twice that. The international partners – NASA, the European Space Agency, the Canadian Space Agency, the Japanese – added their vitally important systems later, so the lifespan of their modules is naturally slightly longer. Therefore, the American segment, as far as I know, is to remain operational until 2030, but there is no guarantee the Russian segment will be actively used by then. It is for that reason that NASA is building a special spacecraft that could independently de-orbit the station without using the resources of the Russian segment.[6]

---

[6] The current de-orbit plans for the ISS at the end of its life, are to use the Russian segment to attach a spacecraft that would steer the station back into the atmosphere at a controlled rate and direct it to a designated area over the South Pacific Ocean. Should the Russian segment, designed to do this, have ceased to function by then, NASA is looking at alternative methods to achieve a safe de-orbit, using a bespoke SpaceX vehicle to dock to a US module.

Below: The finished International Space Station in orbit with its 16 solar arrays fully unfurled and backlit by the Sun. The structure spans 109 meters, equal to the length of an American football field.

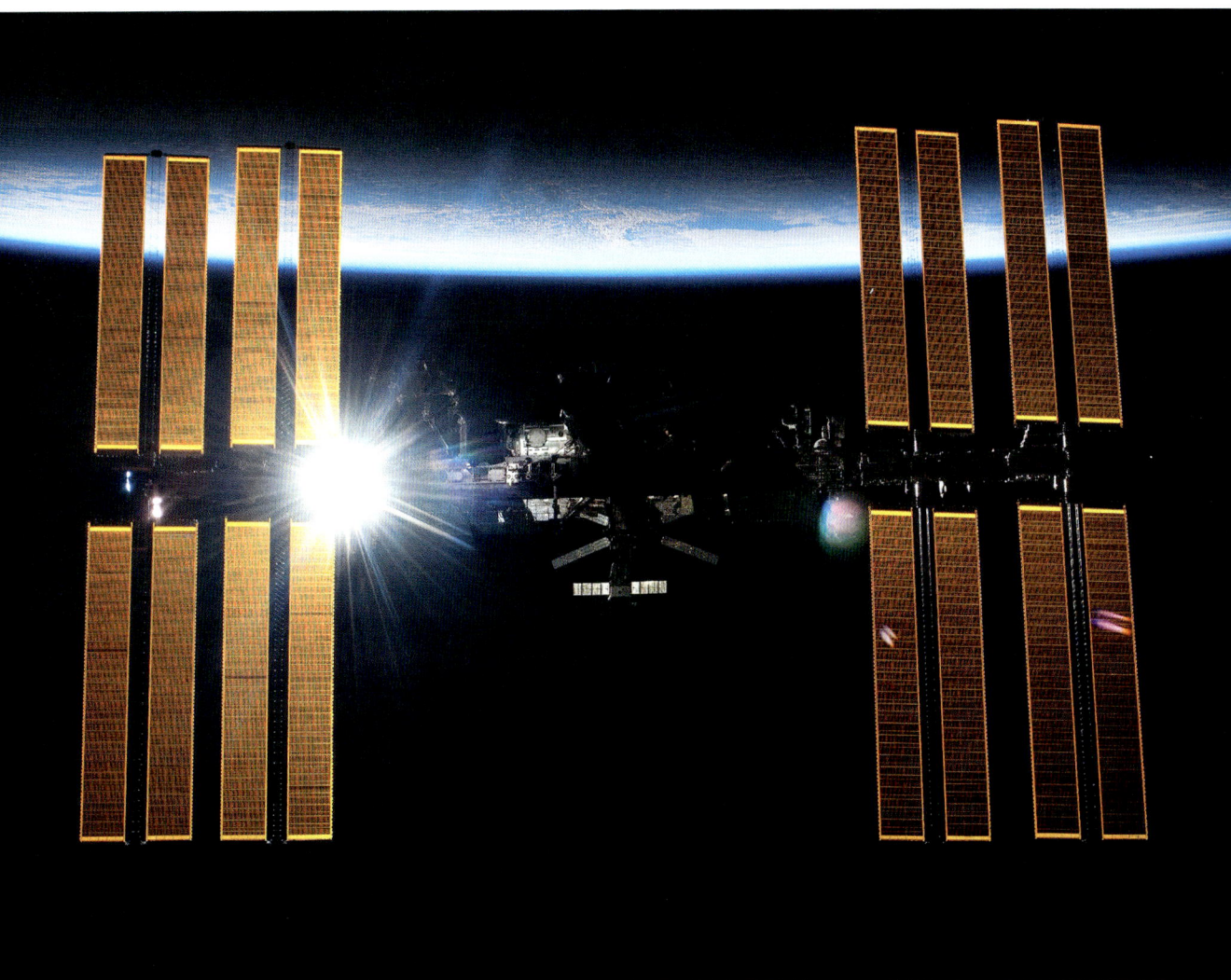

TERRY VIRTS: The metal walls of the Russian segment of the space station are literally cracking, which is really dangerous – NASA folks I've talked to have told me that this is the top safety concern. The cracks in the Russian segment are a kind of a metaphor for what's happening with our relationship on Earth. There are cracks, and we're trying to hold it together, but it's not going to last forever.

In my personal opinion, the decision to continue cooperating with the Russians in space today is the moral equivalent of doing an Arctic expedition or some kind of exploration project with Germans in the 1940s. I'm sure there were good, well-intentioned Germans. I'm sure there was good science to be done. But you just don't do that with a country that is embarking on a war in Europe. And, unfortunately, that's the reality of where we are today.

It's really tragic the damage that Putin's ridiculous war has done to the space station programme. All of the relationships I had with Russians have ended except for one. There's one cosmonaut who I've stayed in touch with – he's a good man – but I had a lot of other relationships, and they have all ended with no contact.

My feelings about my NASA career, which was all about the space station, are mixed. America and NASA built an amazing space station and formed an amazing partnership. I had a really small role in that, but I'm proud of it. I think the space station was very valuable. We've had decades of peace between the USA and Russia. I don't know that the space station was entirely responsible for that, but it played a part.

SERGEI ZALYOTIN: Mir flew for 15 years. The ISS has been up for 25, so it's like an elderly grandmother or grandfather. We need to move forward and build new stations, whether these are going to be purely American, purely Russian or – if the political climate allows it – joint stations.

I hope that in the future we can be successful together. It would be foolish to say otherwise. Our world is too small and too fragile for us to continue these political divisions. Bad relations between our countries did not end our cooperation in space, which is a good thing. I think the warming of these relations should also start from space. If an expedition to Mars takes place over the next 10 to 15 years, God willing, it will be a joint one. It's a symbol we should all pray for.

Right: Cosmonauts Oleg Kononenko and Sergei Prokopiev inspect the Soyuz MS-09 spacecraft that was to transport Alexander, Sergei and NASA astronaut Serena Auñón-Chancellor back to Earth.

Below: NASA astronaut Jessica Watkins, part of SpaceX's Crew-4 mission, looks out of the Cupola on the ISS in April 2022. Watkins was the first Black woman to conduct a long-duration mission on the ISS during Expeditions 67/68. She has a PhD in geology and is one of the NASA's Artemis astronauts, selected to return to the Moon on a future lunar landing mission.

GINGER KERRICK: The first word that pops into my mind [when I think of the ISS] is 'unity', but it doesn't seem enough to cover all the different aspects of what this programme has given me and given everyone on Earth. When you look at all the events occurring in the world today – discord, disagreements, wars – the one thing that would make all those go away is unity. We wouldn't be disagreeing with each other if we had a unified approach. We wouldn't be at war with each other if we felt like we were one. So, maybe unity is the right word. It just seems so simple, but it is hard to achieve unity in the human race, unfortunately. However, there's one amazing example of it that floats above us every day without fail.

The ISS was nominated for the Nobel Peace Prize. It didn't win, but it should have. Most people don't get it. They think it's this thing that NASA created, and it goes round and round, and it's of no benefit. That's why I get so excited when people want to hear our story. In 2050, when we're living in a world in which everyone has hopefully learned how to get along better with each other – maybe we're living on the Moon, or maybe we're living on a different planet – we will look back and the legacy of the space station, as great as it is today, will be even greater.

THE FUTURE OF THE TROPOSPHERE PROGRAMME

JEAN-PATRICE KEKA: I've already got a schedule in mind for the launch of Troposphere 6. We got really close to launching in 2023, and many media outlets and big agencies wrote asking to be invited to the launch. We're opening our doors to everyone who wants to come and see how, in a continent like Africa, we're launching a rocket that can go up to 200 kilometres. Even the president and prime minister will be there. It will be a major event for the country.

Troposphere 5 was a two-stage rocket. With Troposphere 6, because we want to go further, we've designed it to be a three-stage rocket. So, having three stages requires more powerful engines than we've used before because we'll be transporting a heavier mass. It changes the electronics and computer design too, to control the propellant mix autonomously. So, there are many more challenges. But we have to overcome all those challenges if we want to do more.

Troposphere 6 follows the same logic as my other rockets. There are no exotic alloys, such as nickel. Simple steel with carbon and such, that's what I have used. I don't need many things. I make my own calculations to adapt what I find at home, what you can find in regular shops or even what you pick up along the way. Ultimately, I'll have designed a simpler rocket, with a ridiculously low cost compared to existing rockets. And that's my governing principle: where things are expensive, you have to look for cheaper means.

The goal of Troposphere 6 is to make us capable of crafting a launcher as soon as possible.[7] Troposphere 6 needs to carry an even heavier load, and, what's more, go into space, making it the first African rocket to achieve that feat.[8] If this launch is successful, it will make history. It will also be the first time an African state launches its own [suborbital] satellite from its own rocket.[9] We'll drop a CanSat, a satellite in a can, up there and start doing tests on it. So, the Troposphere 6 launch will be very meaningful for us. It'll be the culmination of more than a decade of effort.

We're trying to create impetus with Troposphere 6. It's like the old cars that worked with crank handles: for the vehicles to work, they needed an external input. We're trying to light a spark to launch the DRC's [space industry], because it's for a whole country now, not just for myself. We assume that after this launch, its success will push the state to do even better than we're currently doing. For a country to stand on its own two feet, it must start somewhere.

It's extremely important that we're able to have similar programmes to other countries, because it will give us a way to utilise our own expertise. If, for instance, our programme depends on another country, that country might have its own priorities, and we would be required to follow their wishes even if we did not have the same priorities. Also, space research

Left: Jean-Patrice Keka's rockets.

is a source, and an infinite one in my opinion, of economic growth. Africa has widespread shortages. To fill that void, I believe it's necessary to excel in the aerospace field. When you do this kind of research, you need engineers, computer scientists, physicists, chemists. All those people need to be trained. When people see that the aim is to develop our country, many engineers will start graduating from the universities, and the country will move forward.

There have been a lot of challenges in the creation of Troposphere 6. We had issues with COVID, [the DRC has been involved in] conflict with neighbouring countries, we struggled to fund the work and a misunderstanding even led to the security services arresting me at one point. But where there's a will, there's a way, and Troposphere is now home, at our launch site in Menkao, waiting for the authorities to give the green light so that we can launch.

The launch will happen. I have trust in my government. However, Troposphere 6 is not just any rocket. It's a suborbital rocket, which could easily be mistaken for a missile. There are fears this could cause alarm within the international community and other states might say, 'Isn't launching a rocket a demonstration of strength?' It's possible that's what our neighbours are thinking, but our rockets are made for space research – they don't have anything to do with the military. We're doing science, end of story.

It doesn't worry me that my technology might have dual use. Every technology is dual use. Take the machete – it can cut meat, and it can cut someone's head off. It's a dual-use technology. The problem isn't the technology, it's how it's used. The same technology can be a weapon or a tool for peace – it all depends on the hand that holds it.

My intentions are to build rockets and satellites to support the advancement of our country and our continent. When they build a successor to the ISS, we'll take part. When places in Africa are facing desertification, we can help with cloud seeding. Our sights are set on development, nothing else.

But the goal of launching satellites is just a small step. We are also continuing to look for ways to transport people. We therefore need reliable vehicles. With five years' intensive work, I could put myself in a capsule and come back alive. I really do believe I'll be the first galaxionaut, because I have to show the way. And to see our country join this select group of spacefaring nations would bring great pride and honour. It's not about my own satisfaction, though – rather, it's about opening up paths for everyone. My legacy would be sharing what I've learned with others so they're able to do what I do, so that it doesn't die with me. Because I don't do it for myself. I do it for my country. I do it for humanity. Nothing's bigger than that.

---

[7] A launcher, or launch vehicle, is a spacecraft capable of carrying payloads, such as satellites or passengers, into space.

[8] The Kármán line, at 100 kilometres above the Earth's surface, is widely accepted as marking the beginning of outer space.

[9] A suborbital rocket is one that reaches outer space but does not go into orbit around the Earth.

NEW ENTRANTS INTO THE SPACE RACE

TERRY VIRTS: Space is really important economically. There are nations that have a presence in space, and there are nations that don't, and you clearly want to be in one that does, because they're richer countries and their citizens live better lives. So, I think developing a space programme is a good thing for up-and-coming nations. I think it's important for them to realise they can't do everything, like launches and satellites and space telescopes. You can't do everything unless you're America or China or Europe. But it's important for countries to find their niche and get involved in space, because there's a multiplicative effect where your efforts in space can really help your economy down here on Earth.

They say my space shuttle mission cost a billion dollars. Well, I didn't open up the payload bay and send a billion dollar bills into outer space. That money was spent on Earth. When you develop satellites, those dollars are spent on Earth in industries that require engineers and scientists and technicians. So, the return on investment from space dollars can be really significant for an economy, especially for a developing nation that might not otherwise have a lot of other needs for electrical engineers or aerospace engineers. India is a good example of that. They're doing amazing things in their space programme, and that can carry over into the communications, aerospace and other industries. The space industry can be the seed corn for your economy.

GINGER KERRICK: China landing on the far side of the Moon signalled that we are involved in a new space race.[10] Not every nation pursues space for noble reasons, and even noble nations pursue space for alternate reasons. There are countries that are unfriendly to the United States that could use the Moon as an advantage to their country, and we need to be prepared to counter that.

---

[10] The Chinese National Space Administration successfully landed on the far side of the Moon on 3 January 2019 as part of their Chang'e 4 mission.

Left: This picture released on 11 January 2019 by the China National Space Administration (CNSA) shows the Yutu-2 Moon rover, taken by the Chang'e-4 lunar lander on the far side of the Moon. On 14 January, weeks after landing the rover on the Moon's far side, the Chinese space agency said that China will seek to establish an international lunar base one day, possibly using 3D printing technology to build facilities.

Right, top: Researchers check the Long March 2F rocket, which will send the Shenzhou 8 spacecraft into orbit after the Tiangong-1 space module, at the Jiuquan Satellite Launch Center in Jiuquan in northwest China's Gansu province, 25 September 2011.

Right, middle: Chinese taikonauts from left, Liu Yang, Jing Haipeng and Liu Wang, wave and walk before a giant portrait of China's first taikonauts Yang Liwei, as they depart for the launch pad for their Shenzhou 9 mission to the Chinese space station Tiangong-1, at the Jiuquan Satellite Launch Center in Jiuquan, China, 16 June 2012.

Right, bottom: Researchers work in the control room of the Chang'e 3 lunar probe at the Beijing Aerospace Flight Control Center, China, 10 December 2013.

SERGEI ZALYOTIN: Economically, things now are not like 50 years ago when the Soviet Union entered the race. China, as a socialist country with 1.4 billion people, has a level of economic development that is in no way inferior to American. China, like America, has carried out an unmanned expedition to Mars, which was an undeniable success in terms of moving the science forward. Unfortunately, Russia could not afford such an expedition. While other countries continued to develop, the 1990s, roughly speaking, have set our development back by at least ten years. Therefore, I am well aware that the two powers that are competing with each other now are China and America.

The winner of this race will dominate the world, because the country that develops further technologically will develop better economically. China is really the country of tomorrow. They set themselves a goal in the 1960s, when they started developing [economically], and they are now beginning to overtake America.

I don't know how they will behave in the future, when they are technically a level above everyone else.

KRISTIN FISHER: I think there's a real chance that China beats the USA [in putting humans back on] the Moon. And what's at stake is more than just bragging rights. It's about who gets the best place on the Moon to control the most resources, and that's important for building a sustainable lunar base, but it's also important for some day being able to use water ice to make rocket fuel[11] to then go on to Mars and other places in deep space. The other thing that's at stake is cultural ideologies and beliefs. Do you want a more capitalist view of the world or a more communist one spreading out into the cosmos?

Over the next few years, I think we are going to see exponential growth in terms of what humanity is capable of, from satellites to people in space to various types of spacecraft. It is going to completely transform how we do things here on Earth, and whatever company or country is best positioned to do that is going to be the winner.

I'm an optimist. I truly believe that there's room for all of us in the second space age. I also believe that you need this competition from the Chinese, or else you wouldn't get the funding from Congress to do the kinds of things that NASA is trying to do right now. Nonetheless, there is a lot at stake.

---

[11] As water contains hydrogen and oxygen (H_2O), burning the hydrogen in the oxygen produces huge amounts of thrust to propel a spacecraft.

Below: Shenzhou 9 is the payload that the Long March 2F/G rocket carries, as it launches from the Jiuquan Satellite Launch Center in Jiuquan, China, 16 June 2012.

GOING TO MARS AND BEYOND

GINGER KERRICK: I hope we get to Mars one day, but from a risk perspective, I would much prefer that we return to the Moon, build a sustainable presence there and prove our capability in a place where I can get my crew members back quickly if things go south. With today's propulsion systems, Mars is at best eight months away, and you'd better set a timer when you land, because if you don't leave at the right moment, it'll take you a year and a half to get back. Are you going to run out of supplies? You can't make everything triple redundant, so what are you going to do if something fails? Where are you going to get your replacement parts from? On board the ISS, it's hard to keep a toilet running for longer than a couple of weeks without sending up new parts. So, you've got to make sure you think through all the risks and have a plan to mitigate those. And if you can't mitigate them all down to a comfortable level, then you've got to know what the risks are going in.

SERGEI ZALYOTIN: As a result of my two space flights, I learned that [Konstantin] Tsiolkovsky was absolutely right when he said that the Earth is our cradle, but sooner or later humanity will break into space and reach towards further worlds.[12] This is my vision of the future too. Naturally, it won't happen tomorrow – this is a much more distant prospect, but it will happen. We are well aware that the life of a planet, like the life of an individual, always comes to an end. Accordingly, if we don't explore other worlds by then, if we can't develop our space technology to board our space buses and ships and fly to the newly discovered and explored planets, we'll be here doomed in our cradle here. Understanding this, even in purely practical terms, means we must invest in developing our space technology.

KRISTIN FISHER: Elon Musk's whole mission is to make life multiplanetary, and the reason he wants to do that is because he believes that at some point there is going to be an extinction-level event, be it an asteroid or a pandemic or nuclear war, and humanity needs a plan B. That if we want to survive as a species, we have to find another place for us to live or else we'll go the way of the dinosaurs. I subscribe to that ideology, and I too feel a burning desire to make life multiplanetary. But for me, it's a bit more than that. I'm very curious about the spiritual side of things. Humanity has this gift of consciousness, and in addition to eventually spreading humans throughout the universe, we're also spreading our gifts. So, whatever the reason is, be it the survival of our species or the spreading of consciousness throughout the cosmos, I believe that space exploration should be a priority for all humankind.

---

[12] Konstantin Tsiolkovsky (1857–1935) is considered to be one of the early pioneers of astronautics and space flight. He is the father of rocketry, having created the rocket equation which even today's rockets operate under.

Below: An illustration of SpaceX's reusable rocket Starship operating on the surface of Mars.

ALEXANDER MISURKIN: If we are talking about human space exploration beyond the solar system, meaning interstellar travel with a crew, we can definitely bring this forward and make it more of a reality if we conduct our space exploration not individually as states but as a united civilisation from a united planet. When two humans find themselves in space, it doesn't make the slightest difference what colour their skin is, because they are from a single human civilisation. To me, this is the only common-sense approach as to how humanity should do anything when it is technologically ready to move beyond planet Earth.

Over 50 years ago, Isaac Asimov wrote that the saddest aspect of our life right now was that our science gathered knowledge faster than our society gathered wisdom. This is indeed a global challenge for us all today. The only way to overcome this is to sit down together, with a sense of trust and respect for each other's interests, so that we can determine what our common challenges are, and focus on how to overcome them. The ISS project provides us with a blueprint of such co-operation that we can follow into the future.

KRISTIN FISHER: I I don't think we know what the real value of space is yet. I think it's a bit of wonder, I think it's a dash of some money from a potentially trillion dollar asteroid that you could perhaps harness some of those minerals or energy of the sun and bring it back to Earth. But I think the true value of space is its ability to unite.

The experience of looking down upon the Earth and realising that we're all just from one planet, the boundaries between nations disappear. And that's one of the reasons that I'm so hopeful that, you know, if we do build bases on Mars, if we do build bases on the Moon, the people that build those bases are going to be looking back at Earth in this sea of emptiness, this sea of black, and realise that that's it. That's our only home and we have to protect it. So I'm hopeful that this competition will lead to bigger and better things, not just for any one country, but for humanity as a whole.

Space is one of those few domains or places that you can travel to that really make you reflect on your oneness with all of humanity. When you see Earth from that vantage point, I think it makes people realise that we're so much more similar than we're different and that the potential for true global human collaboration is possible if we can just set some of those relatively small differences aside.

Maybe I'm being naive, maybe I'll be proven wrong, and of course conflict and contests are always a risk, and they may happen. But fundamentally I think that's part of the reason that I'm so optimistic for what this new era in space can bring.

Right: The iconic Earthrise image, taken from Apollo 8, the first manned mission to the Moon, which entered lunar orbit on 24 December 1968. This image of the world became a reminder of the Earth's fragility and isolation; a reminder to protect our precious planet.

Overleaf: A long-duration photograph of the Large Magellanic Cloud, a satellite galaxy to our own, seen through the stars of the Milky Way. This was taken from the ISS as it orbited 260 miles above the Pacific Ocean off the coast of Mexico.